HEROES & VILLAINS

Inside the Minds of the Greatest Warriors in History

HEROES & VILLAINS

Inside the Minds of the Greatest Warriors in History

FRANK McLYNN

This book is published to accompany the television series entitled
Heroes & Villains, first broadcast on BBC1 in 2007.

10 9 8 7 6 5 4 3 2 1

Published in 2007 by BBC Books, an imprint of Ebury Publishing,
A Random House Group Company.

The Random House Group Limited Reg. No. 954009

Addresses for companies within the Random House Group can be found at
www.randomhouse.co.uk

A CIP catalogue record for this book is available from the British Library.

ISBN 978 1 846 07240 6

The Random House Group Limited supports The Forest Stewardship Council
(FSC), the leading international forest certification organisation. All our titles
that are printed on Greenpeace approved FSC certified paper carry the FSC logo.
Our paper procurement policy can be found at www.rbooks.co.uk/environment

Commissioning editor: Martin Redfern
Project editor: Eleanor Maxfield
Copy-editor: Helen Armitage
Designer: Martin Hendry
Picture researcher: Sarah Hopper
Production controller: Antony Heller

Printed and bound in England by Clays Ltd, St Ives PLC

Contents

For Professor Roger Kirby,
without whom this book
would not have been possible

Introduction

The greatest warriors in history

THE RESPONSE OF NORMAL people to the great warriors of history is bound to be ambivalent. On the one hand, we admire their skill, ingenuity and brilliance. On the other, we worry about the human cost of their achievements. A leading scholar of Chinese language and history once told me he could never become interested in the Mongols, as their main contribution to the story of mankind was a mountain of skulls. Another common, but deeply erroneous, prejudice is that 'military intelligence' is an oxymoron. We can admire painters, poets, composers, writers and even statesmen, but never soldiers. But there is a genuine problem here, because the mentality and psychology of warriors is bound to differentiate them from most other humans. One is reminded of the famous exchange between F. Scott Fitzgerald and Ernest Hemingway. Seeking to probe the essential psychology of the multimillionaire, a puzzled Fitzgerald remarked: 'The rich are not like us,' to which Hemingway replied: 'No, they have more money.'

Hemingway's blunt pragmatism is usually applauded, but his answer was actually an evasion of the serious point his friend raised. Something similar applies to the world of the warrior. To dismiss the great captains of history as mere butchers is one-dimensional, but to say wherein their great quality consists is more difficult. In seeking to answer this question, it would be easy to concentrate on Europe, or at best on Eurasia, for this has been the home of all the most legendary warriors: Alexander the Great, Hannibal, Julius Caesar, Tamerlane, Subudei. Our approach in this book is different. We have explored over two thousand years of history and dealt with cultures ranging from Mexico to Japan. Whatever other faults the volume may contain, it is hoped that parochialism is not one of them.

Trying to penetrate the minds of history's great leaders is something that can be done only gradually and with great patience. But if asked to pin down one essential prerequisite for all successful warriors, I would reply that it is an extraordinary capacity for dealing with simultaneous and accumulated stress. The great captains had to deal with conflicts in society and the outer world, with other people and often within themselves. Spartacus, the leader of a slave revolt, not only had to fight the Romans but also to handle opposition from his own commanders, with treachery from renegades, with desertion by his allies (particularly the Cilician pirates) and especially with the doubts that must constantly have been in his mind about the feasibility of his great revolt. Attila the Hun had to deal politically with two very different halves of a divided Roman empire, with traitors and multiple assassination attempts and even strong opposition from his own brother, while negotiating the labyrinth of European politics in the fifth century AD. As king and ruler of an empire, Richard the Lionheart should have been in better case, but on the Third Crusade he faced not only Saladin and the Saracens but also enmity and endemic factionalism in the crusading army, nativist opposition from Christian rulers in Palestine and, most of all, vehement hatred and intrigue in Europe, both from his treacherous brother John and from other European monarchs who conspired to kidnap him. Cortés had to defeat the Aztecs while not alienating his

Indian allies, especially the all-important Tlaxcalans, had to humour the Church, which did not always approve of his barbarism, deal with threats to his position from other Spanish grandees both inside and outside his army of conquest, and suffer the trauma of seeing dear comrades haled off to human sacrifice on Mexican altars. Tokugawa Ieyasu battled not just with his rivals for supreme mastery in Japan but also with Jesuits, recalcitrant Buddhist priests, the imperial court and even his own wayward son. Most of all, he had to wrestle with internal demons that plagued him ever since, at the age of seven, he saw his own father beheaded. Napoleon, while defeating the British and a raft of allies, easily swatted aside intrigues from his own side, loathing of old acquaintances from Corsica, plus envy and sibling rivalry from his brothers. But on top of this, he had to deal with his own 'divided self' – a sensibility torn between mathematical rationalism on the one hand and romantic dreaming and fantasy on the other. When it comes to warriors, the oldest wisdom of all is still valid. We do not necessarily forgive all when we understand all, but we do see them as human beings, both brilliantly talented and deeply flawed.

CHAPTER 1

Spartacus

The gladiator who brought Rome to its knees

SPARTACUS IS UNIQUE among the warriors of the ancient world, and of most other worlds as well. As Karl Marx said: 'Spartacus emerges as one of the best characters in the whole of ancient history.' Most of the great captains of history achieved what they did for power, money, fame and fortune. Alexander the Great said that if he ran out of worlds to conquer he would compete against himself. Egotism, in a word, is usually the key to military achievement. But Spartacus fought to rid himself of a slave's shackles, to be a free man and to enable the other wretched of the earth to be free of the Roman yoke too. This is why Voltaire famously stated that Spartacus's rebellion was 'a just war, indeed the only just war in history'. And that too is why Spartacus has been such an inspiration down the ages. The great conquerors emulate each other. Julius Caesar is said to have wept at the age of 33, when he reflected that at that age Alexander had already conquered the known world while he himself had barely started on his career. Napoleon too dreamed of conquests in the East that would put his name on a par with Alexander's. Yet those who turn to Spartacus are those who are prepared to die for others: Toussaint L'Ouverture in Haiti, Nat Turner in the American

South, John Brown at Harper's Ferry, Che Guevara in Bolivia, all of whom perished, like Spartacus, in a bid to end outright slavery or its close cousin, serfdom.

ROME: A STATE BUILT ON SLAVERY

By the first century BC Rome was the dominant power in the Mediterranean. Already the master of Italy, Spain, Greece and North Africa, the Roman republic and its successor the Roman empire would soon add Britain, Gaul (France), modern Turkey, Egypt, the Near East and Palestine to its domains. By the outbreak of Spartacus's revolt (73 BC) it was a ruthless military state devoted to war and conquest, its cruelty outstripping anything that had gone before when Persians, Greeks and Macedonians were the great Mediterranean powers. Most of all, Rome was a slave state: the peninsula of Italy contained perhaps 6 million people at the time of Spartacus, of whom 2 million were slaves. Entire states built on slavery, as opposed to serfdom, are actually not that common in history: the best-known other examples are the American South before the Civil War of 1861–5 and the empire of Brazil before 1889.

Slavery became a social fetish in Rome, and the ruling class there competed for numbers of slaves. Roman aristocrats prided themselves on their vast slave holdings, which were thought to be the most reliable index of wealth after land itself. The property qualification to be a member of the Senate or ruling body of Rome was 1 million sesterces, but most senators possessed far more than that. It was estimated that if you were worth 3 million sesterces you would also own 400 slaves. Apuleius, the author of *The Golden Ass*, married a woman who brought in 600 slaves as part of her dowry. One multimillionaire left in his will 60 million sesterces in cash and 4116 slaves. Although for legal purposes a slave was considered

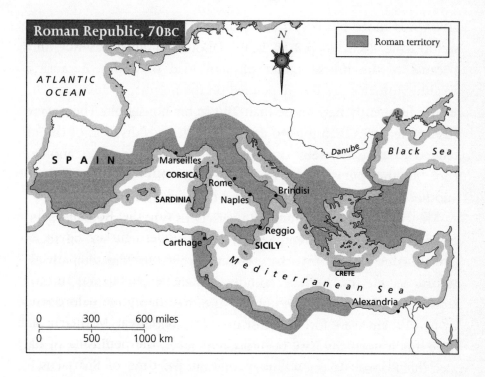

Roman Republic, 70BC

Roman territory

ATLANTIC
OCEAN

SPAIN
Marseilles
CORSICA
Rome
SARDINIA
Naples
Brindisi
Reggio
Carthage
SICILY
CRETE
Mediterranean Sea
Alexandria

Danube   Black Sea

N

| 0 | 300 | 600 miles |
| 0 | 500 | 1000 km |

to be worth 2000 sesterces, most of them could be purchased for an outlay in the hundreds; but the highest price ever recorded for a slave was 700,000 sesterces.

Roman slavery was a complex business, but the society of Rome was unlike that of the other well-known slave states in history and is perhaps best understood by reference to the tribal slavery of Africa as discovered by the Victorian explorers. Whereas in the American South slavery and advanced technology progressed together, in Rome slavery acted as a brake on technology, so that in more than a thousand years of history technical progress was very limited; why discover and innovate and be ingenious if one could easily extract an economic surplus from slaves? There were three main kinds of slaves in Rome and Italy: those who worked in households; rural slaves who worked in the fields and chained slaves on the great estates (latifundia) and in the mines. Naturally the lot of domestic

13

slaves was more privileged and pampered than that of their more unfortunate brethren put to heavy labour, but this was not only because of the much lighter physical load of duties. Because a wealthy Roman's prestige depended on the number of his slaves, he would frequently have more than 200 in his household. There were thus more slaves than jobs to do, and a ludicrous division of labour tended to arise whereby one slave would buy groceries, another cook, another put on his master's shoes, another dress him and yet another follow him around to attend to his every need.

The city slaves were found in a variety of jobs: as street cleaners, builders of baths and temples, factory workers, navvies on roads and aqueducts, as shopworkers, cooks, barbers, hairdressers, nurses, tutors, secretaries, butlers, laundrywomen, housecleaners, seamstresses, schoolteachers. Because of the oversupply of male slaves, many of them were found performing household chores that at the time might normally have been done by females. It will be appreciated that Roman slavery was very different from that in the antebellum American South. Roman slaves were often given positions of great responsibility and with a real prospect of ultimate freedom. Some slaves in large households had *de facto* control of great wealth, supervised other workers (both free and slave) and (here the similarity to African slavery was marked) were themselves served by other slaves. Yet, despite their managerial powers and often sumptuous lifestyle, they owned nothing legally and were subject to the arbitrary whims of their masters; an aristocrat could beat a slave to death with immunity. There were severe penalties for any slave who ran away, including being branded on the forehead. There was always intense hostility between owners and slaves (and vice versa). Hatred was only just under the surface, and the tension was a pervasive motif in instructive literature such as the works of Aesop, the famous slave responsible for the fables.

Where did all the slaves come from? Under the Roman republic most of them were acquired as the result of wars of conquest, but large numbers were also bought from pirates, who at that time infested the Mediterranean. The slave population, then, consisted primarily of defeated peoples, the children of slaves and the products of the slave trade. Further slaves could be bought from outside the Roman domains, from Germany, say, or Parthia in the east. In addition there were abandoned children, and in some provinces destitute peasants sold their offspring into slavery; there were even some races, such as the Phrygians, who did not consider slavery dishonourable.

But slave numbers always posed a problem in Rome, as slaves failed to reproduce themselves. Many of the reasons were obvious. Three-quarters of household slaves were male, and these were the only kinds of slave allowed to have female partners; those employed in agriculture, mining, dockwork, porterage, transport and so on were not. On the other hand, many female slaves were set free (manumitted) so as to be marriageable and bear free children – for a low birth rate in general was one of the perennial headaches for the rulers of Rome; emancipation of these nubile slaves removed the very individuals essential for the maintenance of slave numbers. Moreover, all the evidence is that female slaves were not particularly prolific. Even free people in Rome found it hard to keep up their population levels, given the menace from war, famine and plague, plus poor diet, medicine, hygiene and obstetrics. The average life expectancy of Romans at birth is a much debated subject, but even if we allow an optimistic figure of 25–30 years, the corresponding figure for slaves must be at most 20.

THE START OF THE SLAVE REVOLTS

There was always a contradiction in Roman society between the oversupply of domestic slaves in Rome and the great villas and the labour shortage in the fields, great agricultural estates and mines. Because slaves represented prestige, owners would not release the redundant ones to ease the rural labour shortage, so other means had to be found. The increasing use of condemned criminals in the mines, for example, released other slaves. Then there was enslavement of foundlings or abandoned children – for child exposure and infanticide were common in the Roman world and children, especially girls, were often abandoned by poor parents. But the differential experience of domestic slaves and those performing heavy labour meant that slave revolts were always handicapped by the unwillingness of the domestics to take part, except in very special circumstances.

Rome had known slave rebellions before: the first recorded one occurred in 198 BC in Setia and Praeneste, south of Rome, and ended with 500 slaves being executed; there was another in Apulia in 188. But the two revolts that really shook Rome both occurred in Sicily. In 135–2 BC two separate slave insurgencies broke out, led by slaves named Eunus and Cleon, who later joined forces; they defeated several Roman armies and gathered a host of 200,000 slaves before being finally defeated; the Romans crucified 20,000 prisoners as a dreadful warning. But 30 years later there was another slave war in Sicily, this time led by a man named Salvius. When this too was predictably defeated after four years of warfare (104–100 BC), the Romans decided to send 1000 of the survivors to train as gladiators; the defeated slaves thwarted their captors by committing mass suicide.

All previous slave revolts were dwarfed by the one that broke out in 73 BC under an ex-gladiator named Spartacus. A native of Thrace and some say either of minor nobility or the scion of a family of nomadic pastoralists, Spartacus is unfortunately little known to us: we do not know what he looked like, how old he was (an educated guess would be about 30 at the time of the rising) or what his opinions were on anything; not a single word of his writings or authentic sayings has come down to us. That he was genuinely Thracian is clear, for the name Spartacus was borne by several 'kings' (that is, chieftains) of the Cimmerian Bosporus, on the Black Sea. Some link Spartacus with the Maedi tribe of the River Strymon, allies of Mithridates Eupator of Bithynia, a warlike monarch and famous enemy of Rome, but this seems too neat, an attempt to link Marx's hero (Spartacus) with the hero of the English poet A. E. Housman ('Mithridates, he died old').

The best authorities suggest that he had served in the Roman army on the frontier as an auxiliary, then deserted and spent some time as a bandit before being captured, enslaved and earmarked for training as a gladiator, as a result of his splendid physique and high intelligence, on which the ancient sources are agreed. There is much that is obscure about his early life, and it is not clear why his wife was allowed to accompany him to Rome when he was first taken there as a slave. The legend says that in Rome a marvellous portent was observed. While he slept, his wife, a Thracian seer and prophetess, observed a snake to come crawling up and wind itself around his face; she interpreted this as a sign heralding great power and fortune. What is securely grounded is that Spartacus was taken for his training to the great gladiatorial school at Capua run by Lentulus Batiatus.

GLADIATORIAL ROME

The origin of the gladiatorial phenomenon in Roman society seems to rest originally on their pagan funeral customs. In the early days of its history Rome practised the human sacrifice of prisoners of war to appease the shades of its fallen warriors, and this superstition – that the dead need to be placated and are satisfied with blood – gradually evolved into gladiatorial combat. Instead of a holocaust of prisoners, it was thought more exciting to pit enemies or slaves against each other in combat; hence the original name for gladiators – rustuari or funeral men. Dying men often made provision in their wills for gladiatorial games to be held in their memory, and the custom became so entrenched that sometimes the Roman public would not allow a burial to take place until the heirs of the deceased had provided for funeral games. The December feast of Saturnalia, celebrated with banquets, gluttonizing and orgies, became the favourite date in the calendar for gladiator shows. Gradually games developed from being a private affair into a state institution: in 105 BC for the first time the two Roman consuls, the highest officials in the state, gave lavish official games. Gladiators soon became part of the fabric of society, used not just in shows but also as private bodyguards and hit men. Admired for their physique and martial valour, gladiators were also feared for the same reason but, most of all, despised as the lowest of the low: a male gladiator had the same social status as a female prostitute.

In the era of Spartacus all gladiators were slaves; later, under the empire, free men, aristocrats and even emperors sometimes fought in the arena. It has been suggested that in the imperial era the high profile given to gladiatorial shows, spectacles and chariot races was a kind of compensation for the loss of political rights. Certainly by the time of Jesus Christ gladiators, while still retaining their low-life status, had become the footballers and rock stars of the day. The Roman historian Tacitus noticed this social trend and deplored it: 'How often will you find anyone who talks about anything else at home? And when you enter the lecture-halls, what else do you hear the young men talking about?' Emotions certainly ran high at the games and could degenerate into full-scale rioting. But by the time of the emperors, Rome had learnt the bitter lesson that Spartacus taught them in 73–1 BC. Elaborate sea fights were fought by gladiators in the flooded Campus Martius but, because of the numbers involved, the emperors always feared a mass break-out. For this reason they stationed double companies of praetorian guards around the arena and had an army of catapult shooters and stone-throwers at hand, ready to take action if the gladiators showed any sign of turning their weapons outwards. To understand the era of Spartacus we must imagine a situation where gladiators had not yet become true 'stars', but where cruelty and sadism were just as much in evidence as under the later empire. The main difference between republican Rome 73 BC and all later eras was that the republican élite were not yet aware of the alarming potential threat they had in their midst. All previous slave revolts had been headed by shepherds or pastoralists.

The gladiatorial school to which Spartacus was taken in Capua was at that time the largest in the country. Already there were various kinds of gladiators and styles of fighting, and at Capua the main method used was the Samnite mode. The Samnites were the people of central-southern Italy who had put up the stiffest resistance to the Roman conquest of Italy and had fought Rome almost to a standstill in fifty years of savage warfare (345–295 BC). When Hannibal invaded Italy in the third century BC, in the climactic phase of the wars between Rome and Carthage for supremacy in the western Mediterranean, the Samnites had turned on Rome and sided with Carthage. When the Italian colonies revolted against Rome in the so-called Social War of 93–1 BC the Samnites were again at the heart of the conspiracy. The gladiators at the Capua school, known as 'Samnites', wore heavy armour, a large oblong shield, a leather (or partly metal) greave on the left leg, a vizored helmet topped with a huge crest of plumes and a sword or lance. Although 'Samnites' were in

19

the majority at the Capua school, they had to be trained to fight against different types of gladiator, of which the principal kinds then were the 'Gauls' and the 'Thracians'. Gauls wore heavy cuirasses while Thracians carried a curved scimitar, a small round or square shield, two greaves and bands of leather round the legs or thighs. It has sometimes been (ingeniously) suggested that Spartacus was only a Thracian in the gladiatorial sense, but the best evidence establishes him as a man genuinely from the country of Thrace who was trained as a Samnite.

The scanty historical evidence does not allow us to chart the evolution of gladiators with year-by-year precision, so that it is impossible to know all the kinds of combat in which Spartacus might have been trained at Capua. We know that Thracians fought Samnites and Gauls, and Gauls fought Samnites, and we also know that at some stage new kinds of gladiator appeared: the 'secutor' or chaser, the 'myrmyrillo' or fish man, with helmets bearing the emblem of a large sea fish, and the 'retiarius' or net man, who fought with net, trident and dagger; his art was to sweep the opponent off his feet and then finish him off with the trident; the net had a cord attached to it so that the retiarius could draw it back if he failed to ensnare his prey at the first throw. The retiarius fought with head uncovered and was usually matched against the secutor, but often the secutor would fight a myrmyrillo, or the myrmyrillo and retiarius would engage in a fish-versus-fisherman contest. The retiarius, however, was not usually considered to be a top-flight gladiator and was given inferior accommodation. Following their mania for introducing regional styles of warfare into the arena, when the Romans first encountered the Parthians in the east they were so impressed by their horsemen in chain-mail armour ('cataphractarii') that they introduced them into the arena. There were also mounted gladiators of a more orthodox type, 'velites' or gladiators who fought on foot only with spears, mounted men who wore vizored helmets without eye-holes and were trained to charge each other blind, two-dagger men who fought unhelmeted like thugs in a tavern and even gladiatorial archers. When Julius Caesar made his brief foray into Britain in 55–4 BC he brought back chariot-fighting gladiators as well.

As a Samnite, Spartacus would have undergone the most rigorous training at Capua. Novice gladiators practised with wooden swords for hours on end, thrusting at a dummy or straw man and at 2-metre (6½-foot) high posts, learning the art of lunging, parrying, riposting and feinting. There were connoisseurs of the gladiatorial art who knew the strokes to admire and applaud – the Roman orator Quintillian once said that the structure of a speech by lawyers was much like the alternating thrusts, parries and feints of a gladia-tor in the arena – though, admittedly, most fans came to the games out of primitive blood-lust and to see 'their' side win rather than out of appreciation of the finer points of swordsmanship. Gladiators became massively strong with the hours of practice and the insistence by their instructors that in training they use far heavier swords than they would actually wield in the arena. Their diet was designed to the same end, with beans and barley considered to be particularly good muscle-building food. They were provided with first-class masseurs and doctors: Galen, the greatest physician of the ancient world, began his career as a doctor in a gladiatorial school. They were taught never to blink when a weapon was brandished in their faces, much as a modern-day movie actor is taught not to blink in front of the camera when the arc lights are switched on. Standards of swordsmanship and discipline were high, and security rigidly imposed: no arms were allowed into the dormitories or living quarters of the school for fear of suicide or break-outs.

IN THE AMPHITHEATRE

Finally the trained gladiators had to face the moment of truth and fight for their lives in the amphitheatre. Under the later Roman empire aristocrats and emperors paid for the games, but in the republic this tended to be the job of individual owners and entrepreneurs in conjunction with the 'lanistae' or trainer–managers – another breed, like the gladiators themselves, regarded as the lowest of the low, and commonly compared to professional informers, panders and pimps. This unholy alliance of wealthy owner and cynical manager – so reminiscent of our football clubs today – sold tickets for shows and encouraged betting on the outcome of the individual contests. The football analogy partly breaks down, however, because the really passionate partisanship was reserved for the chariot races in the Circus Maximus in Rome and similar provincial circuits; only the Blues and the Greens of the circus could provide fans with ongoing focuses for frenzied support. One could follow a gladiator only so long before, inevitably, he ran into a better fighter. The giver of gladiatorial shows traditionally threw a huge party the day before the games, at which the gladiators were encouraged to eat and drink to their hearts' content. More thoughtful fighters would brood on their possible death and simply pick at their food, while Greek swordsmen had a habit of taking a dignified farewell of their fellows. Gladiators from Gaul or Thrace had a reputation for insensitivity, for gorging greedily on the groaning board and taking no thought for the morrow.

The games themselves began with second-rate or novice gladiators performing bloodless warming-up exercises. Then there was a second stage where fighters called 'lusorii' fought properly but with wooden swords. Finally came the serious blood-letting. Gladiators were chosen by lot and the real gory spectacle began. Whether the vanquished survived depended entirely on the whim of the crowd, and even emperors thought twice before defying the vocally expressed wish of a bloodthirsty mob. Those who fought valiantly but were defeated were usually given the thumbs up provided they appeared fearless at the prospect of death; but any sign of cowardice, faltering or not fighting to the limit of one's ability would see the mob baying for blood. Finally when all the combats were over, lesser slaves would emerge to rake over the bloodstained sand and haul the dead away. Capua had a huge amphitheatre that could hold 20,000 people, its foundations dug deep into the earth with a low frontage, and with external steps leading up to the top of the auditorium, all constructed over a natural hollow with heaped-up earth around the circumference. Outside Rome, Capua, Pompeii and a few other sites, amphitheatres tended to be wooden and often collapsed with huge loss of life.

In the amphitheatre at Capua, Spartacus learnt his bloody trade, becoming both skilled in arms and immensely versatile in the art of the warrior. Although gladiators usually fought one-to-one, sometimes the organizers of games would decide to exhibit a mass combat and charge higher entrance fees. In general the richer was the provider of games, the more lavish the spectacle. To celebrate his final triumph against all enemies, Julius Caesar staged a battle in the arena with 500 infantry, 30 cavalry and 20 elephants on either side. We may infer that, having survived long enough in the Capua barracks that he emerged as a ringleader of the eventual revolt, Spartacus had high prestige among his fellows and was a veteran of many different kinds of combat, in terms of sheer, physical fighting ability possibly the greatest commander ever. The miasma of cruelty and bloodshed does not seem to have turned him into a psychopath – we are told that he had a moral sense and, when done a favour by someone, showed gratitude. He was also brave and enduring, since he had resisted the route of suicide taken by so many gladiators. Whether he was allowed marital visits by his wife or simply succumbed to the homo-social atmosphere of the barracks we have no means of knowing. The historical novelist Grassic Gibbon portrayed him as bisexual, but this is pure speculation, a likely rationalization of two well-known facts: that the cult of physical perfection and body beautiful among gladiators had an obvious homosexual element, and that women, even some aristocratic Roman matrons, threw themselves at gladiators. Almost the only well-grounded tradition from his time in Capua was the story that Spartacus once defeated his friend Crixus in the arena and then refused to kill him.

THE GREAT ESCAPE

Some time in the year 73 BC Spartacus, Crixus and others devised an involved plot for a great escape that would convulse the Roman world: over 200 highly trained men would stage a break-out and then make a dash for liberty. As was to be expected with a plan involving so many people, the attempt was betrayed. Exemplary punishment should then have followed but, it has been speculated, Lentulus Batiatus was preparing his gladiators for the great Roman Games that took place on 4 September each year. In other words, it is likely that the only reason the entire conspiracy was not nipped in the bud and a handful of ringleaders executed was because greedy entrepreneurs feared they would lose their profits on the games. Assuming the inference is correct, the time of the break-out would be August 73 BC , which makes sense for, with harvests being gathered in the countryside, food would be plentiful. Spartacus and the inner circle of conspirators evidently decided to go anyway; they simply reduced the scale of the operation to about 70 men (at least this is what the historians Plutarch and Appian say, though Florus, a good source, speaks of only 30, while Cicero, the famous Roman orator, said that Spartacus had only 50 men with him at the beginning). Although the school's guards made sure there were no weapons in the barracks, they had not thought through the possibility that the gladiators could arm themselves with other lethal implements, which is what happened. Equipped with kitchen knives, cleavers and cooking skewers, the escape party dispatched the guards and reached the outside of the building, where they found a wagon full of weapons. Even in a gladiatorial school, this sounds like a very lucky find, so perhaps the break-out had been long prepared, and the gladiators had willing assistants on the outside.

Now properly armed, the gladiators headed for the 1300-metre (4000-foot) Mount Vesuvius. In the shadow of the great volcano, they elected three leaders: Spartacus; and Crixus and Oenosaurus (who were both Gauls). The escapees were mainly Gauls and Thracians but there were also some Germans. At first the Roman authorities did not take the escape very seriously: there had been jail breaks before but, sooner or later, the convicts had been hunted down and crucified. Thinking this needed no more than a large-scale police action, the local magistrates at first sent out local militias, which were heavily defeated. Spartacus and his men then stripped the armour from the slain and provided themselves with proper weapons and equipment; they hated and despised the gladiators' swords and spears with which they had fought and jettisoned them at the first opportunity. The defeat of the militia brought Spartacus his first converts, and the augmented slave army camped high up on Vesuvius, on a semi-circular ridge called Mount Somma, about 600 metres (2000 feet) up or halfway to the true summit. This time the Romans took the matter rather more seriously and sent out a praetor named Gaius Claudius Gaber with 3000 troops. Praetors were second only to the consuls in rank: there were six of them every year, holding office as magistrates, civil judges, commanders, governors of provinces and legislators; they were a rank above the quaestors, who performed administrative and financial functions. Guffawing at the supposed stupidity and military naivety of the slaves, Claudius placed a guard post on the single narrow, winding road that led up the mountain. The gladiators, it seemed, were now trapped, since the only other way down was to jump off steep precipices.

Either Spartacus had made an initial mistake and then brilliantly extricated himself from a corner or, more likely, he had baited a trap for the Romans. While Claudius and his men relaxed at the foot of Vesuvius, confident that hunger would soon drive the slaves to surrender or make an impossible assault down the winding road, Spartacus showed the brilliance that would make him the bane of Rome for two years. Mount Somma was heavily overgrown with wild vines, so Spartacus ordered the sturdiest of these climbing plants to be cut down and then plaited and interlaced to make strong ladders. When enough of these of sufficient length had been woven together, Spartacus and his men simply abseiled down the steep cliffs to the plain at the foot of the mountain, leaving one man on the ridge. Once they were all safely down, he threw down all their weapons, then climbed down after them. The gladiators then wound round the side of Vesuvius, surrounded the over-confident, carousing Romans and took them completely by surprise; many were killed but even more fled in panic. This exploit so infuriated the Romans that later historians tried to make it sound like luck rather than planning. But it may be that Spartacus's feat was even more impressive than it seemed, if he really did break out with only 30–50 collaborators rather than 70. Plutarch liked to portray Spartacus growing into the role as the rebellion proceeded, almost as if he could not credit that a mere slave could have planned and then carried out such an ingenious enterprise.

Spartacus was now joined by his wife but, more importantly, large numbers of slaves and even 'free peasants' (actually serfs in all but name) began to flock in to join him. Many of those who joined him after the victory at the foot of Vesuvius knew the region backwards;

household slaves told Spartacus's men where valuables were hidden in houses – and often where their cowering masters lay hidden. Local peasants taught the slaves unusual crafts, such as how to make woven baskets from branches, and suggested taming wild horses so that the gladiatorial army could have a cavalry arm. As the level of expertise in the fledgling army grew, the rebels learnt how to make rough shields out of vine branches covered with animal hides, and swords and spears by melting down the leg-irons and other chains worn by absconding slaves. Slaves from the latifundia and the mines were particularly valuable recruits because they had been honed in a life of toil. Once again Roman capitalism had given too many hostages to fortune. On the latifundia, where a new market-oriented agriculture was practised, chained slaves worked in groups of 12 under an overseer, confined in primitive dormitories at night and always manacled; Columella, a famous Roman writer on agronomy, said it was a mistake ever to remove leg-irons from slaves on a latifundium. These slaves, dehumanized by their brutal treatment, were all the more to be feared when Spartacus freed them.

Even more valuable recruits were the slaves who acted as herdsmen, swineherds and shepherds – the very men who had been the prime movers in the earlier slave revolts in Sicily. Throughout history it has been noticed that pastoral societies produce a warrior tradition, because the basic source of wealth – flocks and herds – is easily stolen and therefore has to be protected by main force. Only the very strongest and physically most powerful slaves could be employed as herdsmen, so in a sense Spartacus was recruiting from the slave elite. Such men had to be able to endure the hardships of the cattle trail, to be agile and sure-footed, and to be able to fight off bandits and wild animals. Many of the herdsmen in southern Italy had been brought from Gaul and even allowed to bring their women with them, but the nexus of allegiance binding them to Roman aristocrats was slender. Yet another invaluable source of manpower for

Spartacus were the 'vilici', bailiffs or farm managers. These men were already used to maintaining discipline so made ideal non-commissioned officers and even lieutenants. Moreover, they had experience in administration and accounting, and knew everything about the area in which they were employed: its resources, supplies and terrain. Finally, they had been recruited originally not from the softer and more effeminate city household slaves but from men with a tough background in back-breaking toil.

Ingenuity and daring had characterized Spartacus's fighting so far and these two factors more and more became the hallmark of his campaign. The Romans' next move was to send out another army under another praetor, Publius Varinius, assisted by two subordinates, Furius and Cossinius, but they had still not taken Spartacus's measure and acted with consummate arrogance, behaving carelessly, failing to post proper guards, going about the task of subjugation in a lazy and indolent manner and − worst fault of all − dividing and subdividing their forces, making themselves an easy prey for a diligent enemy. Publius Varinius seems to have left the actual fighting to his two underlings, for we learn that Spartacus fell upon Furius suddenly and slew 2000 Romans in a surprise attack. Cossinius proved no more able than his colleague: first he was surprised by the gladiators while bathing in a spring in the grounds of a villa at Salinae and narrowly escaped and then, a short while later, was slain when Spartacus stormed his camp. The survivors of both the Furius and Cossinius débâcles compounded their disgrace by mutinying and refusing to fight such dangerous enemies. The lethargic Varinius then finally took the field himself with 4000 reliable troops. Proceeding cautiously, he pitched camp, fortified it with walls and a moat, and waited for Spartacus to take the bait and launch a foolhardy frontal attack. Spartacus easily identified the danger and refused to be tempted. Camped close to Varinius's impregnable camp, he made the Romans believe an attack was imminent by showering them with stones and yelling insults and imprecations. Then he stole away at night, leaving a watchman where the Roman sentries could see him, keeping all fires lit and even manning his trenches with corpses tied to stakes so that they appeared to be courageous defenders. Next morning the Romans missed the usual volley of stones and barrage of insults. Varinius cautiously sent his scouts forward and they soon brought back word that the bird had flown. Thinking he had the gladiators on the run, Varinius abandoned his fortified camp and set off in pursuit, just as Spartacus had guessed he would. He blundered into a well-sprung ambush, and was forced to abandon his horse and his lictors, the men who carried the *fasces* – the symbols of a magistrate's authority. The campaigning season of 73 BC ended with the might of Rome humiliated, the symbols of her power in the hands of slaves and the prestige of the Senate diminished.

SPARTACUS AS ROBIN HOOD

The Romans decided that an all-out effort would have to be made to wrap up this embarrassing slave revolt by the end of 72 BC. The measure of their anxiety was that, almost unprecedentedly, they directed both consuls to enter the field against the slave army; the last time that had happened in Italy against an external enemy was when Hannibal was on the rampage, nearly two hundred years earlier. The two consuls for 72, Lucius Gellius Publicola and Gnaeus Cornelius Lentulus, were given all the powers and men necessary to destroy Spartacus. His brilliant run of successes had meanwhile generated its own problems. So many slaves, peasants and shepherds had flocked to his banner that by now he commanded an army of 70,000 men, many of them armed only with scythes or machetes. For a time all of the extreme south of Italy – Campania, Lucania, Brutium – passed from Roman control into his hands.

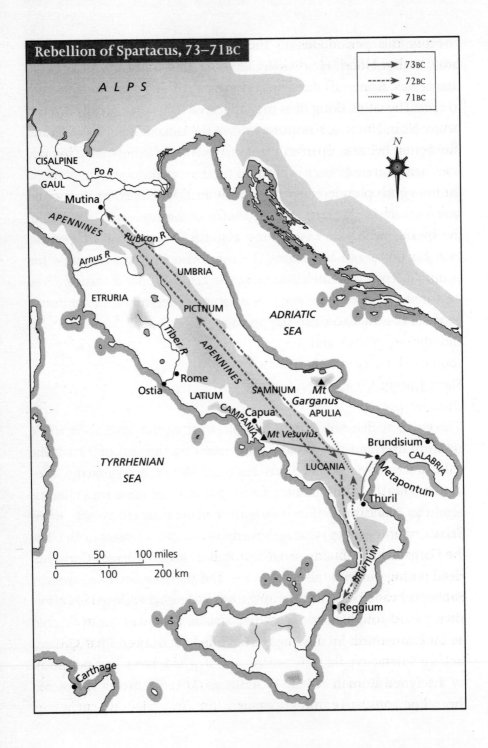

Rebellion of Spartacus, 73–71 BC

→ 73 BC
--→ 72 BC
···→ 71 BC

ALPS

N

CISALPINE
GAUL
Po R
Mutina
APENNINES
Rubicon R
Arnus R
UMBRIA
ETRURIA
PICTNUM
Tiber R
APENNINES
ADRIATIC
SEA
Rome
Ostia
LATIUM
SAMNIUM
Mt
Garganus
APULIA
CAMPANIA
Capua
Mt Vesuvius
TYRRHENIAN
SEA
LUCANIA
Brundisium
CALABRIA
Metapontum
Thuril
BRUTTIUM
Reggium

0   50   100 miles
0   100   200 km

Carthage

From this period derives the legend of Spartacus as a sort of proto-Robin Hood. He divided all spoils meticulously and equally; there was none of the usual division of loot proportionately to rank. The cities along the heel and toe of Italy particularly bore the brunt: Nola, Nuceria, Metapontum were all sacked. The port of Thurii did better because Spartacus needed it as a trading post; foreign merchants carried on a thriving trade, exchanging iron and copper for the goods plundered by the gladiators. With the huge amount of metal purchased, Spartacus gradually improved his armament; the spears, swords and lances manufactured were not at all inferior to the weapons of the Roman legions. His well-known prejudice against the precious metals and their corrupting effect first came to the fore here. Pliny the Elder, who liked to draw moralizing contrasts between the ostentatious wealth of men like Marcus Licinius Crassus and the simplicity and austerity of the early Romans, pointedly contrasted the asceticism of Spartacus with the oriental decadence of Mark Antony, Cleopatra's lover, mocking Antony for using a golden chamber-pot.

Feeding such a host was beyond the resources both of Spartacus's commissariat and the surplus produced on the farms of southern Italy. It was decided that the slave army should split up, in order to live off the land more easily. Spartacus took half the army south, to Brundisium (Brindisi), Consentia and Metapontum, while Crixus, commanding the second division, headed northeast into the Garganus mountains, overlooking the promontory of Garganus Head jutting out into the Adriatic Sea. The Romans saw their chance. Publicola headed east, caught up with Crixus and destroyed his army after a hard-fought battle in which Crixus acquitted himself well; he and two-thirds of his men were killed. It appears that Crixus, lacking Spartacus's flair, was lured out of well-entrenched positions by a feigned Roman retreat. Hoping to do the same to Spartacus, the other consul, Lentulus, pursued him and tried an encircling movement but Spartacus, finding the hinge between the two sectors of the Roman army, instead defeated him.

With the remnants of his force, Lentulus managed to join Publicola and the reunited force then tried to bar the way north to the slave army, when Spartacus abruptly changed his line of march. It seems that Lentulus attempted an encircling movement, leaving an exit between the two armies to tempt Spartacus to walk into the trap, when the ring would be closed. Spartacus spotted the manoeuvre and instead attacked Lentulus at the other end of the 'circle', breaking through so as to separate the two forces. Fragmentary sources speak of Lentulus forming a double line on a hilltop, which implies that the gladiators had to charge uphill to vanquish them. Spartacus had meanwhile left part of his force to pin Publicola's legions; having defeated Lentulus, he then turned back to finish off Publicola. If we can trust the scanty sources for this engagement, it looks as though Spartacus pulled off the 'centre' stratagem – splitting an army with superior numbers so as to have local superiority and then destroying it piecemeal. The details may be obscure but the result was not: once again the gladiatorial army had brushed the Romans aside; they then seized their supplies and held on towards the Apennines.

Spartacus was now heading due north by forced marches. Any euphoria was tempered by the realization that discipline was becoming an increasing problem in his unwieldy army. Things had started to unravel in the far south of Italy, when his men began looting and raping, thus destroying the cooperation of the local peasantry on which any irregular army had to depend. Scholars have often speculated why Spartacus suddenly headed due north, having previously preached the merits of the cattle ranches of the south, but the reason was almost certainly that he felt that only by forced marches could he control his army. If he tarried too long in any one area the ravishings and burnings began, and he was powerless to stop them.

It may be that the epidemic of rape, accompanied often by murder and mindless destruction, was more the work of the 'free' peasantry than the slaves – an inference strengthened by the experience of the slave wars in Sicily 60 years before. On that occasion the rebel slaves, knowing the value of farm produce – not least because they themselves had sweated to generate it – did not destroy goods, crops, harvests and villas but showed a healthy respect for the results of farming and honest toil. It was the peasants tied to the soil by serfdom and debt-bondage who had committed the worst outrages, gutted buildings and destroyed crops for no reason other than rage and envy. Spartacus tried to inspire his men with nobler ideals, and we may conjecture from his forbidding all gold and silver in his camp that he was trying to move his followers away from the love of loot and the corrupt Roman love of luxury. In his failure to sway them lay the seeds of his ultimate failure. But it would be a mistake to see Spartacus as an anachronistic 'noble savage'. Brutalized by his experiences, he could be just as cruel as his enemy and, when news of Crixus's defeat and death was brought to him, he had 3000 Roman prisoners sacrificed to appease Crixus's vengeful ghost.

THE SAVAGERY OF ANCIENT WARFARE

Modern notions of humanist liberalism are out of place when considering the savagery of ancient warfare and especially slave revolts, thought by the Romans to be an affront to nature itself. History teaches us that the worst atrocities, the most cruel barbarities and the most sickening revenge punishments always occur when it is not just a case of one sovereign nation–state fighting another but some more visceral form of combat, such as civil war. Worst of all are those conflicts when at least one side (and often both of them) regards the other as subhuman and as people beyond the pale of

33

civilization: we can observe this syndrome in the attitude of Whig politicians towards the Highlanders in the 1745 Jacobite rising, the French bourgeoisie towards the Paris Communards in 1871 and the Nazis towards the Russians in the war of 1941–5. Similar attitudes prevailed during the Spartacus revolt. The Romans loathed and detested the slave troops the way some people hate snakes, and the hatred was reciprocated. Spartacus liked to stage gladiatorial games in honour of his fallen warriors and ordered his prisoners to fight like gladiators around the funeral pyres of his mourned heroes; on one occasion he killed off 400 prisoners in this way. From a pragmatic point of view, it was a good way of disposing of unwanted mouths in a context where the rebels themselves were often short of food, and it should be pointed out that far more 'civilized' conquerors than Spartacus and the Romans did likewise, Napoleon and Richard the Lionheart being cases in point. In defence of Spartacus it can be said that he had learnt reflex cruelty from his Roman masters, who were adepts at all forms of torture, murder and mutilation. They liked to relax after dinner with private gladiatorial shows, where a cut throat or severed artery would be treated like a digestive drink and greeted with delight. A man named Vedius Pollio, in the time of Augustus, used to throw slaves who had annoyed him into a fishpond full of huge, ravenous lampreys.

The worst refinement of cruelty was the punishment Rome meted out to slaves who rebelled – the fate Spartacus's men knew they would face if defeated. The Romans used four main types of capital punishment: beheading (the most merciful), throwing criminals to the wild beasts in the arena, burning at the stake and crucifixion (the most shameful punishment, reserved for foreigners and, especially, slaves). The Romans got the idea of crucifixion from the Persian practice of impalement, which had been passed on to the Carthaginians by their Phoenician forebears and then to Rome. But whereas in the East the stake alone was used, the victim's arms being nailed above his head or tied behind his back, the Romans added a horizontal traverse on which the arms were extended and then nailed. A vertical pole was set in the ground and a horizontal beam added to make the cross look like a capital T; the idea that they intersected, to form the now familiar Christian shape of the crucifix, was a second-century Christian invention.

Crucifixion was a hideous and excruciating torture, going way beyond most normal notions of cruel and unusual punishment, and there is no authenticated historical case where the victim died in less than 48 hours. Death usually occurred as the result of slow tetanic cramps brought on by spasmodic muscular contractions; the cramps began in the striated muscles of the forearm, then extended into the whole arm, the upper body, the abdomen and the legs. Meanwhile the position of the body impeded the circulation of the blood, producing, to use the medical terminology, progressive carboxyhemia and asystolic heart block. Death was extremely slow, as the contraction of the muscles and the enforced immobility placed an enormous strain on the heart; the pulse slowed and the blood stagnated in the capillary vessels. The heart could no longer eliminate waste matter, the muscles went into spasm and the blood ceased to circulate and carried less and less oxygen to the lungs, being increasingly contaminated by carbon dioxide. All medical research suggests that the victim would feel he was suffocating.

AN INVINCIBLE SLAVE ARMY?

With so few sources, it is impossible to be certain of Spartacus's exact route north, but it seems likely that his army skirted the Apennines on the east, following the narrow coastal strip between the mountains and the Adriatic, passing near the modern locations of Pescara, Ancona and San Marino. Near Mutina (today's Modena), the governor of Cisalpine Gaul, Gaius Cassius Longinus, made a spirited attempt to intercept the invincible slave army but met the same fate as the two consuls; he himself was either killed or narrowly escaped with his life (the sources do not agree). As Spartacus's army headed towards the Alps, freedom was in sight. Once across the frontier, Thracians, Gauls and Germans could disperse to their homelands and their life of thraldom would be behind them for ever. At this very moment, unaccountably, the slave army abruptly turned south and headed back into Italy. This was the most controversial moment in the entire Spartacus rising, and its meaning has divided interpreters ever since. Some have speculated that the gladiators could not feel secure even back in their fatherlands, fearing that the Romans would swear vengeance and that the long arm of the Senate could reach even as far as the Rhine or the Danube. Plutarch thought that the gladiators proper in the army, especially the Gauls and the Thracians, sincerely did want to disperse to their countries but were outvoted or overawed by the Italian majority, who may even have threatened to kill them if they did not turn back with them and continue to lend their expertise. Certainly by this time there was a large majority of Italians in the army, and their motivation was loot. They had no interest in leaving the peninsula and the cold, unknown regions of Gaul and Thrace held no attractions for them.

Reluctantly Spartacus gave the order to turn round and march south. What was in his mind and what was his strategy? In his heart he must have known that the best was now behind them, that he was engaged in a quest forlorn. It is impossible for us to conjecture what Spartacus knew of history, but there must have been those among his 70,000 followers who could have enlightened him on the feats of Hannibal. Of all the warriors who had ever opposed Rome since its rise, Hannibal of Carthage had come closest to success, and he achieved it by a 15-year campaign roving through the Italian countryside, just as Spartacus proposed to do now.

When Hannibal left North Africa on his great exploit in the year 220 BC, the Romans had been as arrogant and complacent as when Spartacus broke out of the Capua school. Proceeding to Spain, Hannibal besieged Saguntum in the southeast of the Iberian peninsula and took it after a long siege. He could not afford to leave such a powerful fortress in his rear and he needed abundant plunder to keep his men (mainly mercenaries from North Africa, Gaul and Spain) happy. There followed the famous passage of the Alps complete with elephants in the summer of 218 BC. On Roman soil finally, he won an initial battle at the Ticinus river, but still the insouciant Romans concentrated on a revolt in Illyria rather than him. When the consul Publius Cornelius Scipio was ordered north with 45,000 men, Hannibal, with just 30,000, destroyed him at Trebia and inflicted 20,000 casualties. Hannibal then marched south and trounced the Romans again at Lake Trasimene in central Italy (June 217 BC); the Romans lost another 15,000 killed and 15,000 captured. There followed a year of the famous Fabian strategy when the Roman dictator Quintus Fabius tried to wear the Carthaginians out by dogging his footsteps but deliberately avoiding battle. It was only when the Romans themselves grew impatient with the slow war of attrition and inactivity that Hannibal was able to achieve his greatest success. At Cannae in southcentral Italy in August 216 he surrounded and annihilated a Roman army 87,000 strong (he had 40,000 in his own force). He killed at least 50,000 Romans at a cost of 6000 dead and 10,000 wounded in his own host. Cannae was, on paper, the most decisive battle in all history and should have given Hannibal complete victory.

Yet the Romans refused to admit defeat. Slowly but surely they gained the upper hand, even though Hannibal continued to beat them in pitched battles. Utilizing seemingly neverending resources, they prevented Hannibal from taking Nola in southern Italy, themselves besieged and defeated his chief Samnite allies in Capua, threw back the invasion of a fresh Carthaginian army at Beneventum, and opened two new fronts, in Spain and Sicily. First they besieged and defeated Hannibal's great ally Syracuse, then they cleared Spain of Carthaginian armies in a long series of battles. Finally they defeated Hannibal's brother Hasdrubal at Metaurus (207 BC) and the new army he was bringing to reinforce Hannibal. Increasingly isolated and with diminishing numbers, Hannibal tried everything to regain the initiative, even marching on Rome itself in 211 BC to try to make the Romans lift the siege of Capua.

Confident that Hannibal lacked the siege engines to take their great city, the Romans shut themselves inside their walls and continued the siege of Capua. Finally, when Hannibal was left with no options in Italy, the Romans played their trump card by sending their ace general Scipio Africanus with a large army to Tunisia, taking the war to Carthage itself. In a panic, the Carthaginian authorities recalled Hannibal to North Africa. Without a proper cavalry arm Hannibal went down to defeat against Scipio at Zama (202 BC), and the cause of Carthage was lost for ever. Here, then, was a clear historical precedent. The most deadly enemy Rome had ever faced had defeated their armies in seven battles, had laid Italy to waste over 16 years but still ended by losing the war. Rome had proved that its resources were inexhaustible, its willpower indomitable, and its confidence such that it could wage four other campaigns even while it contained Hannibal, widely acknowledged to be the greatest military genius of all time after Alexander the Great. What hope was there that Spartacus could do any better?

A UNIQUE CONJUNCTURE

The optimists in Spartacus's camp urged the unique conjuncture of historical events in the year 72 BC. When Rome faced Hannibal, the

city was united in a common purpose. But now it was riven by factionalism, the republic was just recovering from a ten-year civil war
between the generals Marius and Sulla, whose followers spent most
of the decade of the 80s killing each other, the Italian allies were
disaffected, as shown by the so-called Social War of 91–88, to say
nothing of the support given their own revolt, and Rome was at this
very moment occupied with three other major campaigns. There was
something in these arguments. The issue of land and the grotesque
inequalities of wealth made Rome a very brittle society indeed, and
there was a state of simmering sub-civil war between the old, aristocratic, patrician faction known as the 'optimates' – basically men
who wanted nothing whatever in the social structure to change –
and the 'populares' – also men of wealth but ones who wanted to
co-opt the common people for their programmes in return for a
minimal amount of land redistribution. Most of the populares were
no more interested in the mass of the people than the optimates but
considered, rightly, that shunning and despising them openly was
bad and dangerous politics. Although Rome was officially ruled
by the Senate, the sovereign body of 600 legislators, in reality the
Senate usually rubberstamped the decisions of the man who had
most power: in the 90s this had been Marius and in the 80s Sulla.
By the time of Spartacus's revolt real power rested with three men:
Lucius Licinius Lucullus, the champion of the optimates, C. Pompeius
Magnus (Pompey), the heir of Sulla, officially an optimate but really
unconcerned about anyone but himself, and the demagogic populist
and plutocrat Marcus Licinius Crassus.

The two great war leaders, Lucullus and Pompey, were already
engaged in wars overseas. In 74 BC Nicomedes IV, king of Bithynia in
Asia, bequeathed his country to Rome, and the Senate sent out the
two consuls Lucius Licinius Lucullus and M. Aurelius Cotta to admin-
ister it. The presence of the Romans on his doorstep inflamed king
Mithridates of Pontus, an old enemy of Rome who fought long wars

against Sulla in the 80s. Mithridates attacked Cotta but was forced to retreat when the more energetic Lucullus cut off his supply line. Lucullus was a good general, but Mithradates was a wily opponent who continued to give him the slip and, when Lucullus defeated him, crossed the border east into the independent kingdom of Armenia, ruled by his friend Tigranes, and defied the Romans to extend the war that far east. Meanwhile in Spain, Pompey was still engaged in a grim struggle against Quintus Sertorius, heir to Marius's mantle, and thus in exactly the same relation to Marius as Pompey had been to Sulla. Sertorius was another talented general, and Pompey yet a third, so the Spanish soil saw a grim to-and-fro struggle that went on for five years (76–1 BC), with neither side gaining a clear advantage.

As if these two wars were not enough, Rome was also involved in a protracted war with the pirates of the Mediterranean, a scourge of the sea lanes for more than a generation. As far back as 78 BC the Senate had appointed P. Servilis to a three-year command aimed at extirpating piracy, but he was notably unsuccessful and was replaced in 74 BC by the seemingly more energetic M. Antonius. Alas for Roman hopes, Antonius was badly beaten by the pirates of Crete in 72 BC. When Spartacus turned back from Gaul, then, Rome was on the back foot in Spain and on the high seas, with the prospect looming of a possible war against Armenia in the east.

Spartacus, though, knew the importance of morale. He realized that if his vast army simply turned back into Italy to loot, it would soon disintegrate as marauding bands broke off to rape and plunder; all discipline would be lost and he himself would be hard put to stamp his own authority, which had already taken a knock when his men refused to follow him over the Alps. He therefore decided to announce that the return to Italy had been part of a grand strategy after all: having shown the Romans that, like Hannibal, they could range at will up and down Italy, he was now going to top Hannibal's exploits by marching on Rome and succeeding in taking it where the Carthaginians had failed. He knew, however, that he would have to humour his volatile men and to watch his back.

THE MARCH SOUTH: AN ULTIMATE STRATEGY?

Despairing of ever beating Sertorius in the field, the Romans eventually offered a bounty of 100 silver talents and 20,000 acres of land to anyone who would deliver him to them, dead or alive. The bait was taken by Sertorius's own officer Marcus Perperna Vento in 72 BC, who murdered Sertorius and thus brought the Spanish war to an end. Such a fate, it was clear, could easily be Spartacus's, and the evidence suggests that he tried to distract his men from any such intrusive thoughts by a programme of constant activity and forced marches. We hear of more Roman prisoners being offered as human sacrifice to appease the gods of war. Then Spartacus set off down the western side of the Apennines, making for Rome, burning everything in his path, killing all prisoners to save feeding them, and progressively slaughtering his pack animals for meat as he went, using the excuse that the beasts of burden were slowing down the rate of march.

But as the army reached the latitude of Rome it suddenly swung away east towards the foothills of the Apennines before bypassing Rome altogether and winging down south to the east of Capua, eventually crossing the trail where the luckless Crixus had left Campania for his last stand in the Garganus mountains. We can only guess at the deliberations that went on in the slave army on the trek south. Had the gladiators' leaders concluded, wisely, that Rome was impregnable except to an army equipped with powerful siege engines and habituated to siegecraft? Did morale lift to the point where Spartacus once more felt secure and resurrected the strategy of quitting Italy? Was there already fatal discord and factionalism in the army?

What is certain is that Spartacus's army eventually began heading for Brundisium (Brindisi) with the avowed intention of making for Sicily and raising the slaves there, where there was a firm and old tradition of servile revolt. Could it be that at the Alps the army had drawn back at fear of the unknown, not relishing the colder northern climate but that now they had the prospect of warm Sicily, well known to them, where they could be among friends and trusted allies, they finally conceded the wisdom of Spartacus's exit strategy? Maybe it had taken another march along the spine of Italy before all the short termism of rape and plunder wore off and they were finally prepared to listen to the wisdom their great leader. For on the face of it there is something more than a little absurd about returning to the place where they had been 12 months before. Either Spartacus had been originally uncertain about his ultimate strategy or, much more likely, he had not been master in his own house and had been forced to go along, however reluctantly, with majority opinion. What is absolutely clear is that Spartacus, his later champions notwithstanding, was never a social revolutionary, an early egalitarian, Jacobin, communist or proletarian leader. He simply wanted to get back to his homeland and the simplicity of his early life. The march on Rome was probably always a gimmick to keep his army together; Spartacus had no pressing interest or business there, and the only later commentators who imagined he could have were those who falsely saw him as an ideologue or one who wanted to transform Roman society root and branch. Such ideas were probably beyond the mindset of even the most advanced and revolutionary thinkers of the ancient world; the Christians, the most original theorists of the Roman era, continued to believe in slavery and keep slaves.

CRASSUS VERSUS THE 'DEMON' SPARTACUS

As the third year of the war approached, even the severe, stoical and long-suffering Romans began to crack; Plutarch speaks of fear and desperation that no one could exorcize the demon Spartacus. Lentulus and Publicola were recalled in disgrace and confined to civil affairs for the rest of their consulate, while the command of Roman armies in Italy was given to Marcus Licinius Crassus, the emerging third man in the struggle for mastery in Rome, after Lucullus and Pompey. Crassus, already a rich man, would become a byword for wealth; 'as rich as Croesus' was the Greek phrase, but the Roman equivalent could well have been 'as rich as Crassus'. Starting life as a wealthy man (with a fortune of 300 talents or 7,200,000 sesterces) but not yet a plutocrat, he built up a treasure hoard of 7100 talents (216,800,000 sesterces by the end of his life).

Aged 42 when the Spartacus rebellion broke out, Crassus made his early money from property speculation in Rome. The city of the republic, and even more so under the empire, was a literal tinder-box, its seven-storey buildings and slum hovels regularly going up in a fiery inferno – the result of drunken carousal, malice, carelessness, litter, primitive heating and total lack of civic pride. Crassus's special-ity was to buy up burning buildings at a knock-down price; the owners, uninsured, facing ruin or the exorbitant costs of rebuilding, were only too happy to retrieve something from their investment. The artfulness of Crassus was that he had built up an expert cadre of building specialists – slaves who specialized in architecture, plumbing, sanitation, brickwork and all the other arts and crafts of homemaking; additionally, he had his own private fire brigade, adept at putting out a fire rapidly once Crassus had struck a deal. Even if a building were completely gutted by fire, Crassus reckoned he could rebuild and resell at a huge profit. Naturally, his enemies whispered that it was Crassus's 'secret army' of slaves who started most of the fires in the first place, and the suspicions seem well grounded.

Although Crassus has come down to us in history as a slum landlord, he was actually too shrewd financially to base his fortune on such a rickety foundation. The uncertainties of the Roman real-estate market, the difficulty in enforcing civil judgements against the penniless, and Rome's floating population meant that it would be foolish for any would-be entrepreneur to tie up capital in rental property. Crassus's speciality is not unknown today: the purchase, rapid refurbishment and quick resale of properties bought for a song. Soon he was diversifying into agriculture – the only respectable form of investment in ancient Rome – mining (especially the silver mines of Rome), tax farming and banking. He used his corps of trained slaves as a private manpower-rental service, hiring them out as readers, copyists, waiters, charioteers – whatever, indeed, the market would bear. He also became a great political boss, dispensing favours and patronage, binding people indissolubly to him with loans, establishing a powerful nexus of clientelage and obligation. Soon he was in a position to challenge Lucullus and Pompey politically, but one thing gnawed at him: his lack of military glory. By their conquests Pompey and Lucullus won prestige that no mere capitalist could ever attain, for Roman culture exalted military prowess above everything else, even wealth. Of course the two went hand in hand, for Crassus found to his consternation that Pompey was soon even richer than he was, despite minimal entrepreneurial talent. The profits from military conquest, and the loot and treasure a victorious general could bring back to Rome, outran the profits even the most inspired capitalist could make. It was time for Crassus to diversify into warfare.

'RICH ENOUGH TO BUY A LEGION'

The recall of Lentulus and Publicola gave Crassus his chance. While the disgraced consuls grubbed around, saving face in the revived office of 'censor' (in charge of electoral registers) while trying to raise money for the wars with dubious and obviously pro-Pompey measures, Crassus put his vast wealth at the disposal of the Senate, always provided he were appointed supreme commander. The Romans liked to debate what it meant to be truly rich, and the later Latin writer Pliny the Elder said you could call no man rich who could not pay for and equip a legion of 5000 men. The Romans were great ones for outdoing each other in hyperbole, and this ancient tradition about being 'rich enough to buy a legion' was soon transmuted by the multiplier principle into 'rich enough to buy an entire army'. Crassus, in a word, offered to equip and pay for six legions, which varied in strength between 5000 and 6000 men. He promised he would enter the field against Spartacus with 40,000 of the best fighting men ever to don Roman uniform. It is worth pausing to consider what he was committing himself to. The six legions would cost 450 talents in annual pay for the infantry, plus 37–38 talents for the cavalry and the same amount for the auxiliaries. Making allowance for higher pay for officers and rounding up all sums, one would arrive at a sum of around 600 talents. If we assume that by the year 72 BC Crassus was already halfway to the 7100 talents he definitely possessed in 55 BC, just before his death, it can be appreciated that he could comfortably afford to run an entire army and still have plenty of loose change.

The first problem for Crassus was to ponder why Spartacus had won so many victories, for on paper it seemed unlikely that a slave army could beat Roman legions, or at least not more than once. The ancient historians' battle descriptions do not allow us to solve this conundrum, but the wealth of material available on Roman methods of fighting and their encounters with other irregular armies, such as the Germans, enables us to make an informed conjecture. Roman legions typically fought in three lines, arranged in a chequerboard formation. In the first line were the 'hastati' (literally 'spearmen'), the youngest troops, and behind them, but standing to one side so as to leave a gap were the 'principes' (the 'principals'); behind them in turn, and aligned so that they were in the same position as the front line, were the reserves or 'triarii'; a Latin saying 'to be back to the triarii' meant you were in a desperate plight or on your beam ends.

The Romans fought in long, thin lines, allowing them to outflank an enemy who used deep formations. In a typical battle, the skirmishers (velites) would first engage the enemy and try to break up his formation and then, when pressed, retreat back through the gaps of the chequerboard; if there had been a continuous line in all three ranks, chaos would have ensued. As the enemy approached, the principes hurled lighter spears over the heads of the first ranks and then the hastati launched the heavier javelin or *pilum*. The front line then drew its swords and engaged in close combat; so great was the impact of clashing armies that legionaries often jumped up on enemy shields to cut downwards. The brilliance of the three-line chequerboard formation was that it allowed the Romans to fight in relays. Because the labour and stress of combat were so colossal, men could fight intensely only for short periods before falling back exhausted, at which point the lines behind would move up and fill the gaps. The rested fighter would then return to the fray, relieving his substitute, and so on. If the battle were going badly, the hastati fell back to the second line and then, in dire straits, both would fall back to the triarii.

In battle timing was everything. If the second line were deployed too soon, it could get snarled up in the frontal fighting and become exhausted; but if deployed too late, it might get swept away in a rout. As the three lines made for a flexible response, it was difficult for a single line of fighters, as in the traditional phalanx, to punch its way through. But if delivered with enough speed and ferocity a single charge might knock the hastati back too fast for the second line to deploy effectively. We know from German campaigns that barbarian armies liked to fight in phalanx formations with overlapping shields or a wedge formation designed to pierce through like a gigantic can-opener. To win against the legions, an enemy usually required an unusual combination of numerical superiority, speed, surprise and favourable terrain.

DIFFERING MOTIVATIONS: FREEDOM AND DECIMATION

Spartacus was brilliant at manoeuvring the Romans on to ground that favoured him, which is surprising since the legions tended to fight only on terrain of their own choosing; if they could not find it, they liked to shelter behind their nearly impregnable fortified camps. The textbook location for a Roman commander to draw up his army was at the top of a hill with a stream below, so that the enemy would have to ford the stream and then charge uphill. If we except one case, where it seems Spartacus did have to charge uphill but still won, he excelled in getting the legions to abandon their usual tactics. Of course one can understand why Roman commanders departed from their own boilerplate when fighting the rebels. Flexibility in tactics was always important, as their great victories in the past had shown, and they were fighting an enemy who did not employ the usual methods; underrating the opponent and sheer arrogance also came into it. Yet perhaps the one winning card Spartacus held was his men's motivation. Ever since Marius abolished the property qualification for serving in the legions, the Roman army was a professional force made up of landless citizens, interested in pensions and the land and cash payments with which a victorious campaign would be rewarded, not with killing the most possible enemies. The gladiators and slaves, on the other hand, fought from desperation, to win loot and to remain free; if they lost, there was no state-sponsored cushion to fall back on. This must sometimes have been the deciding factor when the two hosts clashed. In the long, traumatic slog of hand-to-hand fighting, stamina and willpower were all important, the energy required to make just one more charge, and one more surge; everyone knew that the side that gave up first would be routed and slaughtered.

Crassus decided that the way to beat Spartacus was not to rush into early battles but to keep his troops trained to a high pitch and await the right opportunity. Basing the bulk of his forces at Picenum and Picentia (Vicenza, near Amalfi), he sent his legate Mummius south with strict orders to ape the example of Fabius against Hannibal: never engage, follow the enemy, but if he advances, immediately withdraw. Naturally, Mummius, a man on the make and inspired by a lust for glory, disobeyed the instructions when he thought he saw a chance for an easy victory. But Spartacus was merely toying with him and inflicted yet another sharp defeat on the Romans. The defeat of Mummius created a sensation, for eyewitnesses attested that his men had panicked when the 'barbarians' surged in among them; those who escaped the carnage were those who had dropped their weapons and fled.

Crassus was just as meticulous about discipline as Spartacus, but whereas the gladiators' leader had to lead by example, cajolery and personal charisma, Crassus had brute power on his side. He decided to revive the ancient Roman practice of decimation, reserved for those who had acted the coward in battle. If an army unit were judged deficient, disobedient or lacking in courage, one soldier in ten was selected by lot and cudgelled to death by his former comrades. It is most likely that 500 men of the defeated army corps were chosen and 50 'ringleaders' then bludgeoned to death. Decimation is one of those practices that, like cannibalism, has lost nothing in the telling but, again like cannibalism, it was an historically real and terrifying ordeal. Decimation was practised in Roman armies often enough not to be considered remarkable, even though it began to die out under the empire – Roman ingenuity found more interesting cruelties to perpetrate. Along with the public execution of prisoners, it taught Roman youth the lesson of what happened to those defeated in war and inculcated the notion that it was better to die honourably in battle than a coward's death later. It was a rough-and-ready system of dinning terror into raw recruits, as the Roman historian Tacitus conceded: 'When every tenth man in a defeated army is beaten with clubs, the lot falls also on the brave. Making an example on a grand scale inevitably involves some injustice. The common good is bought with individual suffering.'

A DESIRE FOR PEACE DENIED

Spartacus soon learnt that the era of fighting against Roman second-raters was at an end. He put out peace feelers to Crassus, which the Roman rejected contemptuously, not even deigning to reply to overtures from a slave barbarian, in his eyes a subhuman. Even if he had been a humanitarian, disposed to take Spartacus's peace offer seriously, politically Crassus had no choice. The mere fact of opening talks would have destroyed his credibility. He would have been disgraced and thus disappointed in achieving all the ends for which he had taken the command in the first place. Besides, he was currently much concerned with image-building. Attempting to be a man of the people by trying the old dodge of learning men's names and calling them by their given name whatever their station, Crassus made a point of inviting ordinary and humble people to his lavish banquets, instead of the oligarchs from the old families, even though he had no real interest in the common man, posing as their saviour merely to build up a political power base. His aristocratic critics accordingly said that Crassus had suspect sympathies, that he was too soft to deal decisively with Spartacus and his 'people's army'. Crassus therefore deliberately set out to prove himself a man of iron; as Plutarch said of him 'he was strong because he was popular and because he was feared – particularly because he was feared'. Crassus was determined to show Spartacus no mercy. Continuing his policy of dogging and then striking at stragglers, rearguards or any slave units that made a mistake, he scored a first great victory by ambushing and annihilating a column of 10,000 men operating independently from Spartacus. These may have left the slave army to forage because of food shortages or, more likely, they were a breakaway group led by men who thought themselves as good as Spartacus and no longer wanted to take his inconvenient orders.

By this time part of the problem for the gladiators was that their escape plans were unravelling. Having originally designed to embark with the Cilician pirate fleet at Brindisi, Spartacus received word from them that the embarkation point had been changed to the Straits of Messina; they did not care to tarry so near where Crassus's army was operating. After agreeing the very high fees demanded by the pirates, Spartacus and his army crossed Lucania to the north of the instep of Italy and reached the sea by the straits near modern Reggia di Calabria. Here a terrible shock awaited them: there were no ships for them. The venal C. Verres, governor of Sicily, who ran the island like a personal fief or (appropriately) as an early mafioso, took alarm when he heard that Spartacus and his dreaded slave army intended to cross to Sicily and whip up the dispossessed and resentful who still remembered the earlier slave revolts. The last thing Verres wanted was Spartacus and then after him, presumably, Crassus interfering with his lucrative scams and schemes of extortion. Verres knew the way to the hearts of the pirates. He simply topped the offer from Spartacus by a huge margin, handed over the gold to the pirates and persuaded them to sail away into the sunset. This action by Verres was a masterstroke and probably doomed Spartacus more effectively than anything else. When Verres was brought to trial in 70 BC for extortion, the prosecuting counsel, Cicero, denied that Verres had had anything to do with the destruction of the gladiators. But Cicero, a sycophantic, snivelling trimmer and pathological liar, as well as great orator, was by this time cosying up to Pompey, whom he identified as the rising star, and was certainly not going to give Verres any credit.

A SLAVE ARMY IN DESPAIR

The despair in the ranks of the slave army can be readily imagined. At first the disoriented fighters tried to turn themselves into sailors and marines and build rafts made of wooden beams, with buoyancy provided by barrels lashed together with vine tendrils, but after several unsuccessful trial launches and many drownings they concluded, rightly, that only stouter, larger craft could make headway against the treacherous currents. The Straits of Messina were notorious for swift-flowing cross-currents and ancient navigators found the passage of these waters so difficult that the legend of Scylla and Charybdis had arisen, whereby whichever side of the straits you tacked, you were in mortal danger, either from a sea monster or an all-devouring whirlpool. Dejectedly, Spartacus took his army further north along the coast of the Tyrrhenian Sea and pitched camp on the promontory of Scyllaeum (near the town of modern Scilla). Now Crassus thought he saw his chance and pounced. Yet another reason for his abandoning caution was that he had just learnt that Pompey, victorious in Spain, was returning to Rome. Crassus knew Pompey's methods of old, knew that the last time Pompey had come back with a conquering army he had even faced down the terrifying dictator Sulla and got his own way about having a triumph. Pompey's return at worst could mean civil war, but even on the best possible reading of the runes he would certainly demand to be made consul for the year 70 BC out of turn. Although on the 'Buggins's turn' principle used at Rome for elite careers Crassus should have been the shoo-in candidate for consul, having completed the required period since being praetor, the danger was that Pompey might have one of his tame minions elected as the other consul. The only way Crassus could secure his own election and make Pompey think twice about using his victorious legions to cut through the political factions at Rome was to win a crushing victory over Spartacus.

THE LEGEND OF CRASSUS'S WALL

Nothing more clearly illustrates the way the Spartacus story has been encrusted with legend than the story of Crassus's wall. Plutarch tells us that Crassus decided to blockade the slave army by building a wall, complete with trench and paling (very like Hadrian's Wall) that would cut the gladiators off from the rest of mainland Italy and starve them out. According to Plutarch, Crassus cut a ditch 4.5 metres (15 feet) wide and 4.5 metres (15 feet) deep across 55 kilometres (35 miles) of the toe of Italy, extending as far as Thurii. Amazingly, most modern historians have accepted this story, even though its implausibility is obvious. Hadrian's Wall, little more than twice as long, took two or three years to build. Moreover, such a line would have served no purpose, as Spartacus would have been left as master of 4150 square kilometres (1600 square miles) of territory, a larger area than the state of Rhode Island, and there would thus have been no possibility of starving him out. A 55 kilometre (35 mile) wall would have left Crassus's army hopelessly thinly spread, so that the slaves would have had no difficulty breaking through. And, finally, such a combined wall and ditch would have involved the displacement of hundreds of thousands of cubic feet of earth and stone, ranging, depending on where we trace the line of the putative wall, between 31,000 and 44,000 cubic metres (1,100,000 and 1,540,000 cubic feet); even today's engineers, with modern technology and machinery, would have found this a daunting task, and an impossible one in the time span required.

Roman historians were obviously impelled to ludicrous exaggeration to show that Rome could rise to any occasion; in psychological terms such hyperbole looks like 'compensation' for the terrible fear Spartacus had instilled in them. The best Roman historian of Spartacus, whose work unfortunately survives only in fragments, Sallust, tells us the truth of the matter. Crassus in fact shut up Spartacus on the promontory of Scyllaeum, an area large enough to accommodate the slave army but small enough to be effectively blockaded. Crassus was able to bottle Spartacus up because the landward side of the promontory is already separated from the surrounding terrain on the east and west by the deep ravines of the Torrente Livorno and the Vallone d'Angelo; a relatively short wall and trench, not the monster described by Plutarch, would easily have provided the cork for the bottle.

Spartacus had missed a trick by not attacking Crassus's men while they were building the wall. Too late he spotted the danger, and now the slave army was in a serious fix, for its provisions were running low and could not be replenished in such a tiny area. Spartacus ordered an assault on the wall, but the Romans easily threw this back, inflicting heavy losses. He then tried probing at various supposed weak points along the line, trying feints and sallies, hurling fiery brands and burning branches into the ditch to try to frighten the defender. Finally he crucified a prisoner in full sight of Crassus's men, partly to try to overawe the Romans but mostly, it is surmised, to warn his men of the fate that awaited them if they did not break through. In desperate straits, he was finally rescued by the winter weather. One snowy night, with snowflakes, slush and sleet vying for elemental superiority, he rushed a poorly defended section of the line and filled the trench with whatever the slaves had to hand – dirt, stones, wood, slaughtered cattle, butchered prisoners – to form a primitive causeway over which the gladiators poured to safety. Spartacus used one-third of his army to make the first attack and then, when the 'causeway' was ready, took the rest across. The Romans dared not stand in their way, knowing that one-third of the enemy was already in their rear, and simply abandoned the wall. Tired, demoralized but glad to have escaped what looked like certain destruction, Spartacus's men made their way to a prearranged mustering point in the woods in the foothills of Mount Sila.

DIVIDED THEY STOOD

Morale in the slave army had now plummeted, partly because they had taken such heavy losses when penned in the Scyllaeum promontory, though the historian Appian's estimate of 12,000 dead is a ludicrous overestimate. Hungry and demoralized, Spartacus agreed to divide his army to make foraging easier and began heading north along the Via Popilia, taking the smaller of the two forces; the larger contingent of Gauls and Germans was commanded by Castus and Cannicus. By the lake of Palo (near the modern town of Buccino), just off the Via Popilia, Castus and Cannicus pitched camp, thinking Crassus could not possibly be close behind them. The Palo lake was about 5 kilometres (3 miles) in circumference and surrounded by mountain slopes; its water level fluctuated dramatically with the seasons, being brackish and undrinkable in the summer but now (in the winter) sweet-tasting from the melting snow in the mountains;

it was to get a good fresh-water supply that Castus and Cannicus had stopped here. But Crassus had urged his men in hot pursuit by forced marches and they now fell on the encamped slaves.

Before ordering his attack, Crassus sent 12 cohorts (6000 men) round the other side of the mountains with orders to climb up and then descend and attack from the other side. Castus and Cannicus were within an ace of annihilation as Crassus gave the signal to attack. His stratagem worked perfectly, and the rebels were thrown into panic and confusion by the sound of ululating voices in their rear as the cohorts sent on the encircling movement began to rush down from the mountains. Battle was joined, but how long it lasted is uncertain, though some sources speak of heavy casualties. Suddenly, as if out of the blue, Spartacus appeared in the Roman rear, and Crassus, fearful of being defeated in the confusion, broke off his assault. Evidently a message had got through to Spartacus from the other army – one version was that two slave women had gone up the mountain to perform their toilette and saw the back-shooting Romans swarming down, enabling the Gauls to send out a plea for help – but his seemingly miraculous appearance in the nick of time did nothing to harm his reputation; truly he seemed to be indispensable, and without him the slaves were lost.

After saving the Gauls and Germans, Spartacus agreed with their leaders that they should continue to operate separately, to ease the food and supply situation; they continued north along the Via Popilia as if heading for Rome while he swung east with his force. The initiative seemed to be passing from Crassus. The Senate, impatient initially with the siege of Scyllaeum and now thoroughly alarmed by Crassus's failure to bring Spartacus to a decisive battle, seemed to have lost its collective nerve, especially now that the slave army was further north than Crassus and on the road to Rome. They sent orders that Pompey was to return from Spain immediately with his legions while Marcus Lucullus, brother of the greater Lucullus still fighting around the Black Sea, should land another army at Brundisium. The order to Lucullus was the result of jockeying by the optimates. Having lost faith in Crassus, and fearing that a victorious Pompey would be over mighty, they called in Lucullus's brother so that they would have an army ready to oppose Pompey if necessary.

The landing of Marcus Lucullus at Brundisium narrowed Spartacus's options, as it precluded a return to the Roman heel – his preferred option. Marching north on the road that branched off from the Via Popilia at Eburum, he followed the Nares river north towards Aquilonia, where there was a junction with the Via Appia. Here he could watch for the approach of Lucullus's army, which would have to come this way from Brundisium. If he were lucky he could meet and defeat Crassus before he could effect a junction with Lucullus. The scenario seemed feasible, as Crassus in turn had been forced to divide his forces, with one section (under his quaestor Tremellius Scrofa and his legate L. Quinctius) following Spartacus while he dogged Castus and Cannicus. Crassus was taking the easier option; even at this stage he had no real stomach for a head-on confrontation with the great slave leader.

Departing from his usual practice of retreating before the Romans and then returning unexpectedly to catch them, as it were, on the counterpunch, Spartacus decided to dig in at an entrenched camp for the expected showdown with the Romans. He made his preparations at the Silarus river (not far from the modern town of Capo Sele) where, 140 years earlier, Hannibal had defeated the Romans. It was now the spring of 71 BC. Spartacus was confident that, if discipline held, he could still defeat Crassus and then turn to face Lucullus; he seems not to have known of Pompey's recall. His confidence in his own abilities was justified, for when Scrofa and Quinctius came up he did to them what he had done to Lentulus and all the others. Once again the Romans were utterly defeated; Scrofa was wounded and barely escaped with his life.

CRASSUS, CONQUEROR OF THE BARBARIAN SLAVES

But meanwhile Castus and Cannicus had once again demonstrated their military incompetence, and this time there was no Spartacus to save them. Crassus planned an elaborate decoy to lure the unwary Gauls to their doom. First he built two camps and threw defensive stockades around them, keeping the trappings of his staff and secretariat in both locations so as to bamboozle the Gauls' spies as to his ultimate intention. While they were distracted by the conundrum of the two camps, Crassus led the bulk of his infantry out and hid them in the mountain foothills of the highlands of Cantina. He then sent out a large cavalry force to offer battle to the Gauls, with orders to draw the enemy after them gradually into the ambush in the foothills. Although the feigned retreat was a well-known tactic, when the Roman cavalry withdrew after being deluged with arrows, Castus and Cannicus decided to follow it, reasoning that such a large force must mean that Crassus was now putting all his bets on his cavalry. At the appropriate moment the cavalry screen suddenly dashed behind the wings of the main force. The trap was sprung and to their horror Castus and Cannicus realized they had blundered into a skilfully laid ambuscade. The slaves fought with their usual valour but were utterly annihilated. As the fruits of his victory Crassus was able to send to Rome five legionary eagles, 26 manipular standards and the *fasces* and axes of five lictors, along with many other spoils, all recaptured from the slaves.

The news of the disaster was brought to Spartacus, who at once decided to abandon his plans for a decisive battle with the Romans in favour of guerrilla warfare. Crassus was only a day's march away and, unless they immediately struck camp and marched away with all speed, he would be on them. But now his men, doubtless flushed with their victory over Crassus's deputies, told Spartacus they were tired of running and would retreat no further. With a heavy heart

Spartacus faced the inevitable. Heavily outnumbered and lacking a cavalry arm to contest the field with Crassus's horses, he must have known his hopes were very slender. In a melodramatic gesture he brought his trusty war horse to the front of his army; this reliable steed was to him what the famous mount Bucephalus had been to Alexander the Great. He drew his sword and then told his men in a loud voice that their situation was now do-or-die: if he won the battle, he could have his pick of horses but if he lost he would not need one. With that he slew the horse on the spot; this was a traditional gesture of war-chiefs before a great battle and would have been understood as such by his soldiers.

The battle that followed was a desperate slugging affair, with both sides fighting for life, the Romans knowing that if they did not perform valiantly, they would have Crassus and his decimation to deal with. Eyewitnesses said that Spartacus tried to cut his way through the throng of fighting men so that he could close with Crassus in personal combat but was severely wounded in the attempt. With a spear-thrust in the thigh, he continued to fight while his strength lasted, finally dropping on one knee but continuing to brandish his sword until the tide of victorious Romans rolled over him. Evidently they did not recognize him for, to Crassus's fury, the body of Spartacus was never recovered, and so was saved the indignities the Romans had certainly prepared for it. Perhaps 5000 to 6000 slave troops had fallen – the Romans lost 1000 dead – but the battle was as decisive as it well could be. Crassus had achieved his ambition.

Although the rebels lost the battle, they were not annihilated. Some contingents escaped and established a base in the country around Thurii, where they turned to banditry and plagued the countryside until finally being extirpated by the father of the man who would become the first Roman emperor, Augustus. Less fortunate was another party of 5000 who headed north to Rome only to run smack into Pompey's legions from Spain. Landing in Rome in

March 71 BC, Pompey immediately set off south to try to win new glory and was lucky enough to be able to pick up these crumbs from Crassus's martial table. Never one to hide his light under a bushel or to perform his patriotic duty by stealth, the arrogant Pompey then wrote to the Senate that although Crassus had beaten the slaves in pitched battle, it was he himself who had finally ended the war. Naturally enough, this infuriated Crassus and sharpened the already fierce rivalry and jealousy between the two rising stars.

But Crassus had, on his own terms, reason enough to feel pleased with himself. In six months, as a private citizen bearing an extraordinary commission from the Senate, he had accomplished what praetors and consuls had not been able to do in more than a year. Unwittingly, Spartacus had acted as the ladder for his ambition. After hunting down the unfortunates who had sought refuge in the mountains nearest the Silarus battlefield, Crassus crucified 6000 of them. The Appian Way from Capua to Rome was lined with crosses bearing the rotting bodies of crucified slaves, one placed every 35 metres (40 yards) along the 200 kilometres (125 miles) to the capital. The Romans thought this was a fitting and just end for those who had attempted to overturn the 'natural order'. Neither Crassus nor any other Roman aristocrat would have felt a scintilla of pity for the defeated, for their entire society and their own privilege rested on a basis of slavery. All regimes rest ultimately on force, but few have ever done so as nakedly as the Roman; according to Roman culture, not to mete out terrible exemplary punishment was merely to encourage slaves to rise again.

ASSESSING THE SPARTACUS REVOLT

Crassus and Pompey came to terms so that they shared the consulate in the year 70. In contrast to Pompey, a victor in a foreign war against a recognized enemy, Crassus, as the conqueror of detested barbarian slaves who did not count as human beings, let alone worthy opponents, did not merit a full triumph. He had to settle for an *ovatio* or acclamation, a lesser distinction, but used his political muscle to get the Senate to grant him a laurel crown, an honour usually reserved for a formal triumph, instead of the myrtle crown customary for the *ovatio*.

During the year of the Crassus–Pompey consulate, something happened that brought the memory of the Spartacus years back forcibly to Roman consciousness. The young Julius Caesar, one of the 20 quaestors that year, was travelling to Rhodes when he was captured and held for ransom by pirates. Once ransomed, he demanded action against the pirates. The Roman governor of Asia was dilatory, so Caesar bought up a coastal defence unit, tracked down his captors and crucified them – a controversial action as he had no legal authority and, as quaestor, had no business pursuing private vengeance. Nevertheless, the incident reminded the Roman authorities that they had not yet settled accounts with the pirates, who could have made the Spartacus revolt even more of a nightmare for them if Verres had not bribed them. In the year 67 BC the Senate gave Pompey plenipotentiary powers to rid the seas of pirates, who by this time were interrupting Rome's all-important corn supply from North Africa. Provided with the most extraordinary resources – 120,000 infantry, 5000 cavalry and 500 ships – Pompey swept the Mediterranean clean, starting in the western seas and then gradually tightening the noose around the pirates in their favourite lairs in the Aegean. Having vanquished the ancient maritime foe, Pompey refrained from crucifixion but settled the marauders on agricultural colonies well inland. By the year 66 BC Pompey had replaced Lucullus in the war against Mithradates and brought that too to a successful conclusion.

By the year 60 Crassus, Pompey and the rising star Julius Caesar had become so powerful that they agreed to rule Rome as a triumvirate under the formal sovereignty of the Senate. They could afford to look back on the Spartacus revolt and comfort themselves with the thought that its defeat had been inevitable. The reality was that the events of 73–1 BC were far more serious than that, and for a long time the Spartacus rising had been a close-run thing. Crassus had the numerical superiority in the final battle where it really mattered, but things could easily have been otherwise, especially if Spartacus had managed to get to Sicily and raise the flag of rebellion there. Spartacus's army in Sicily, a densely populated island, disaffected, with a history of rebellion against Rome and with ample food supplies, one of the granaries of the empire, would have been a tough nut to crack. At least eight legions would have been needed for a successful invasion and, with so many troops absent from the Italian peninsula, the temptation for Samnites and other slaves to take their chances would have been enormous. But a Sicilian adventure required both the cooperation of the pirates and a monolithic slave army.

The inability to get even his own men to present a united front was Spartacus's downfall, as it has been in history for so many popular movements, but maybe a political genius (which Spartacus was not) could have achieved it. If, further, he had been able to make common cause with the pirates against Rome, matters would have been very serious indeed. On the other hand, the experience of Hannibal suggests that, at this stage in its history, Rome was probably invincible and its resources inexhaustible. The most likely solution to the 'what if?' scenario of a Sicilian campaign is what historians call a second-order counterfactual. In other words, the slave war would probably have lasted another ten years, but in the end Rome would still have prevailed. In the long run Roman adaptability and flexibility, the discipline, organization and logistics of its armies, their longer combat effectiveness and *esprit de corps*, the fact that the troops could win land, citizenship and cash pay-outs and, perhaps most of all, the fact that Rome had an infinite capacity to absorb and replace manpower losses over time, would surely have been decisive.

IRREGULAR ARMIES VERSUS THE NATION—STATE

History shows that armies of irregulars have little chance when ranged against the apparatus of a nation–state unless they go for a quick kill. The Jacobite Rising of 1745 provides many instructive pointers for the Spartacus revolt and many direct analogies. Both rebellions began in August and ended in April but both faltered when its leaders failed to attack the enemy capitals (respectively Rome and London), for a rapid coup d'état was the only way forward. Irregulars, like the Scottish Highlanders or the gladiatorial army, can win stunning victories against professional troops in the short run by using shock tactics like the massed charge, but in the end lack of supplies, money and materiel stack the odds massively in favour of the state apparatus. A nation–state can, if necessary, raise a mercenary army, can raise loans – and even with the primitive state of Roman banking the vast estates of the élite provided unlimited collateral – can sustain itself through crisis after crisis while the irregulars, dependent on spoils, a precarious food supply and an ever shrinking army are almost bound to lose.

Roman historians taunted Spartacus with the failure of city slaves to rise for him, just as Whig propagandists in the eighteenth century jeered at the Jacobites for their failure to win mass support. The writer Arthur Koestler in his novel on Spartacus even compared the lack of 'spontaneous revolution' elsewhere in Italy with the fallacy of Lenin's 'socialism in one country'. But most humans are not heroes and, knowing the ferocious backlash if they back the losing side, remain quiescent until the winners emerge clearly. Spartacus understood this aspect of human nature, as we can infer from his well-known refusal to accept deserters from the Roman army into his own force. Quite apart from the natural suspicion that such men might be double agents, Spartacus reasoned that deserters were egotists who would fight for a cause only as long as it suited their interests; such natural turncoats and backsliders were best not dealt with in the first place.

In military terms, the response to Spartacus as a warrior has always been ambivalent. On the one hand are those who cannot believe that a 'mere' slave could defeat the might of Rome and speculate, quite unnecessarily, that Spartacus must have had high rank in the Roman army, possibly as a centurion. In fact, as a veteran of every type of combat, single and group, in the arena, he would already have been familiar with most kinds of tactics and the Roman military mentality. As a general, Spartacus has received a bad press, doubtless because professional historians so disliked the cult of Spartacus as revolutionary hero. As against Marx, who called him 'a great general (unlike Garibaldi), a noble character and a genuine representative of the ancient proletariat', modern critics have argued that Spartacus did not beat very much in military terms, despite his nine victories over the Romans; to use the vernacular, he fought only against mugs. It is often pointed out that once Crassus faced a credible military opponent, he was found wanting. His campaign against the Parthians in 53 BC, waged to try to achieve military parity with Pompey and Caesar, ended with one of Republican Rome's worst defeats, at Carrhae, and his own ignominious death.

It is true that in 73–1 BC Rome's first-class troops were with Lucullus in Asia and the second-rank ones with Pompey in Spain, leaving only third-raters to deal with the slave revolt. But against this Spartacus had to lead a polyglot, multicultural army. He could not use battle tactics that required uniformity of language but had to use signs and symbols instead. He had to use his personal prestige and charisma to charm and cajole; he could not simply issue an order and back it up with the threat of decimation. Spartacus may not have been one of the great captains of history, but he was a better general than his critics allege, and it is hard to imagine that he would have made the stupid and arrogant mistakes his old enemy Crassus made in the land of the Tigris and the Euphrates.

CHAPTER 2

Attila the Hun

The Mafioso-warrior

ATTILA THE HUN, in direct and indirect ways, finished off the Roman empire, which had lasted more than a thousand years but by the early fifth century AD was in terminal decline. The empire reached its zenith in the second century AD but thereafter had been wracked by a number of crises, caused by plague, the decline in the production of precious metals, heavy taxation, a shrinking agricultural base and even manpower shortages. Historians still debate about the fundamental causes of the decline of Rome, and a German scholar in the 1980s famously listed 600 different factors that were relevant. Yet it was undoubtedly the invasion of 'barbarians' on Rome's northern frontiers that gave the *coup de grâce* to the empire. Germanic tribes, impelled by a number of different motives – overpopulation, envy, relative deprivation, agricultural crisis – spilled over into the empire and could field huge and formidable armies if the Romans dared to oppose them. Many of the tribal names have become famous for one reason or another – Alans, Burgundians, Vandals and especially Goths, who divided into a western branch (Visigoths) and an eastern one (Ostrogoths).

At the very moment these 'barbarians' were pressing on the

Roman frontier, the empire itself was imploding, and by the beginning of the fifth century, it had split into two, with two emperors, one based at Constantinople (modern Istanbul), the other in Italy. The division of the empire had been prefigured as far back as 286 when the emperor Diocletian introduced the 'tetrarchy' – basically splitting the empire into four with four emperors. This was supposed to make it easier to respond to barbarian threats in the different parts of the empire, but simply led to fragmentation, with none of the emperors living in Rome. The emperor Constantine reunited the empire under one emperor but then, as the fourth century wore on, the empire more and more divided into two. The emperor Theodosius (379–95) was the last to rule a united Roman empire. Having no confidence in his sons, he bequeathed the west to Honorius and the east to Arcadius. All of Theodosius's talent and ability went to his daughter Galla Placidia, one of the great female figures of the Ancient World. In 395 the division between east and west became permanent. Henceforth there was a western empire based on Italy, and with nominal control over North Africa, Britain and parts of France and Spain, and an eastern empire centred on Constantinople, still controlling Asia Minor (modern Turkey), Thrace (modern Bulgaria), Egypt and large parts of the Middle East. But the German tribes were already well entrenched in France and Spain and poised to take over completely.

A NEW SECT OF BARBARIANS

Attila the Hun is one of those rare historical phenomena that appear suddenly like a thunderbolt, cause terrific havoc then fade away as rapidly as they first appeared. This is what makes the discussion of his case so difficult. The Huns themselves are historically problematical, as they existed as a significant factor for a hundred years at most. Nothing is known of them before the fourth century AD: attempts to identify them with the Hsiung-nu nomads who devastated the frontiers of China at an earlier date are unconvincing. Real historical evidence for the story of the Huns is almost nil. They themselves were illiterate and ignorant of their early history. They minted no coins and, as they were nomads, archaeological evidence is tenuous. All that can be said for certain is that they came out of that notorious limbo, 'central Asia'. It was the year 376 when they first impinged on the West, for in that year the Roman empire learnt that a new sept of barbarians, the Huns, had conquered the Romans' hitherto most formidable enemies across the Danube, the Goths and the Alans. Rome had particularly feared the Ostrogoth kingdom, a confederation of German tribes stretching from the rivers Don to the Dniester and from the Black Sea to the Pripet marshes. Now it turned out that the Goths had been annihilated and their king (Ermanarich) had committed suicide. The Huns' defeat of the Alans was even more complete. Having vanquished the eastern Goths (the Ostrogoths), the Huns went on to assail their western cousins, the Visigoths. Shunted forward by this sudden assault, the panic-stricken Visigoths crossed the Danube in large numbers. Some 200,000 of them descended on Adrianople, routed a Roman army and overran the Roman province of Pannonia. It is thought that Hunnish cavalry, recruited by the Visigoths, provided the firepower that made Adrianople (376) such a signal defeat for the Romans.

In 395 the Huns crossed the frozen Danube and invaded Thrace and Dalmatia. At the same time they invaded the eastern empire and devastated the areas of Cappadocia, Syria and Armenia. But for the moment they concentrated on the western empire: in 405 they entered northern Italy, in 406 they crossed the Rhine into Gaul (France) and in 408 attacked the Danube provinces of lower Moesia. There were occasional forays into the eastern empire: a raid by the first known Hunnish king, Uldin, forced Constantinople to complete the building of city walls on the land side in the year 413. Part of the extreme complexity of this period of history is that there were no real superpowers, but very many powerful kingdoms and mini-empires that appeared from time to time to play a part in the drama. In 415–20 it was the turn of the powerful Persian empire, stretching away eastwards to Afghanistan from the rivers Tigris and Euphrates to experience the wrath of the Huns, but on this occasion the Huns were badly beaten and withdrew. In 420 Persia and the eastern Roman empire went to war; the Huns used the opportunity to raid Thrace once more. In the early fifth century the Huns were never the implacable enemies of the Romans and were sometimes even friends and allies. In the succession struggles in both parts of the empire to decide who would be the next emperor, both the east and the west used Huns as mercenaries; sometimes civil war pitted east Romans against west Romans and the Huns would be engaged to take part in those conflicts also. At this stage the Huns were raiders, lacking any overall objective or strategic purpose, high-grade technicians of warfare who were happy to hire out their skills to the highest bidder.

The Huns were the first of the mounted nomads from central Asia who would terrify the world, in a sense forerunners of the later and much more fearsome Mongols. As nomadic pastoralists they were hunter–gatherers and did not practise agriculture. Legend speaks of 'hordes' of Huns, by which we imagine hundreds of thousands of warriors, but the truth is that the manpower base of the Huns was always limited; they should be numbered in thousands and at most tens of thousands rather than the myriads of mythical imagination. Population size is always determined by the availability of food, and a very large area of pasture land was needed to sustain even a small number of Huns and their horses; we should think of many small groups crisscrossing the steppes of central Asia in the early days.

Because of problems of food and pasture, the average Hun raiding party was only about 1200 strong.

The building-blocks of Hun society are easy to identify. Each family lived in a tent, between six and ten tents made up a camp, and several camps in turn made up a clan – the basic social unit of Hun society. There were ten clans to a tribe, so a tribe was maybe 5000 strong (which could convert into 1200 warriors), but the tribe always had to split into camps of 50 or so people to seek pasture and forage. Huns had a reputation for faithlessness over treaties, but the Romans made the mistake of thinking of them as a normal nation–state. The problem was that a treaty signed with one tribe of 5000 Huns in no way bound other tribes. Famine was a constant feature of early Hun society, and there was no economic surplus that would allow a leisured class of nobles to emerge. Nor was there any domestic slavery; the only slaves Huns had were those captured on campaign. The Huns had leaders only in times of war, and the leader was whoever had the greatest military reputation. It was a peculiarity that all discussions and councils were held on horseback.

THE FEARSOME HUNS

Why did the Huns have such a terrifying reputation, far eclipsing that of the Goths or other barbarians? In part it was because their numbers were so wildly overestimated. Because of their speed and mobility, the mounted Huns could cover vast distances in a short time, so there seemed to be thousands of them everywhere; often it was the same tribe reappearing in different locales. Their skill with horses and their amazing horsemanship, which made them seem like Centaurs, contributed to a reputation for not-quite-human efficiency. The Romans feared them because the Huns appealed so mightily to the many slaves, bandits and other fifth-columnists within the Roman empire, who hated their masters and overlords and looked on the Huns as saviours. Then there were the terrifying stories about the Huns brought back from the frontier: it was said that they sacrificed to the gods their old people, and even their own parents, so as not to have useless mouths to feed. But some part of the unreasoning terror engendered by the Huns must be ascribed to their physical appearance. Romans found them loathsome to look at and thought them as ugly as gargoyles; both their facial appearance and their marmot-skin clothes frightened 'civilized' Italians who expected even the enemy to dress and look like themselves.

Yet what frightened the Romans most was the Huns' astonishing skill with the bow. An arrow shot from the heavy bows the Huns used – which took exceptional strength and years of practice to master – at a range of 45–90 metres (50–100 yards) could have more penetration than many modern bullets. The Hunnish mounted archers could fire an arrow every two seconds and the best of them could shoot 1000 arrows in the course of a 12-hour battle. The Huns used huge horse-drawn wagons to bring hundreds of thousands of shafts to each battle – each wagon contained about 100,000. Their archers would fire three salvoes, the first from 140 metres' (150 yards') distance, the next at 90 metres' (100 yards') range, and the final shot from 45 metres (50 yards). It was estimated that the Hunnic cavalry could fire 1000 arrows in the first five seconds, and another 1000 in the next five, and 12,000 arrows a minute had the killing power of ten machine guns. An attacking Hun army could deluge a 10-metre (11-yard) front with 50,000 arrows in ten minutes. Moreover, the Huns' nomadic lifestyle meant that they did not need to go into winter quarters and could fight all year round, though the Romans thought that the time to attack them was February–March, when their horses had been weakened by winter.

Before Attila there was, however, no permanent Hunnish confederacy, so that after a campaign the Huns tended to split into tribes, clans and camps. The great confederacy of the 370s that laid low both the Alans and the Goths then disintegrated. The Hunnish threat was staccato, discontinuous and intermittent, which was why bursts of campaigning fury were followed by long fallow periods; to put it another way, the natural tendency among the Huns was always towards entropy. Uldin, for instance, did not lead all the Huns but only a section, and this tendency continued until Attila's time. But Uldin did begin a process that was important in the history of the Huns, for he forced the Ostrogoths to become, in effect, his serfs.

The first crucial event in the history of the Huns was the move from the Aral and Caspian Sea to the lands west of the Black Sea; formerly they had dominated the Alans to the east of the Black Sea, but from the beginning of the fifth century they had the Ostrogoths in the Ukraine as their underlings. The second was when they left these regions and made their base on the plains of Hungary. The Ostrogoths were crucial to the Huns' supremacy, being both good fighters and expert agriculturalists. So, on the one hand the Huns solved their problem of scarce numbers by getting the Ostrogoths to do the infantry fighting for them, while they appeared as the élite force of mounted archers, ready to deliver the crucial shock blow that would win battles. On the other, they forced the Ostrogoths to work for them in the fields, extracting a surplus both to feed their armies and to allow a leisured class of noble warriors to appear. By the early 420s the Huns were showing signs of being able to put together a lasting confederacy that would make them a permanent and deadly threat to both halves of the Roman empire. The brothers Rua and Octar ruled jointly over a powerful league of Huns but, as so often in Hunnic history, one of the joint rulers conveniently disappeared. In 432 Octar died, leaving Rua as sole ruler.

ATTILA AND FLAVIUS AETIUS

Octar and Rua had another brother named Mundiuch, lower in the pecking order. Mundiuch's importance was that he was the father of two future Hun rulers, Attila and Bleda. Almost nothing is known of Attila's childhood, though it is conjectured he was born in 406. The one thing that is known is that in his adolescence he had a Roman friend who would later be his nemesis, and this in itself is a fascinating story. Flavius Aetius was born in 396. Called by later historians 'the last of the Romans', Aetius was a man born out of his time, an idealist who cherished the ancient glory of Rome but had to make pragmatic decisions in a very different world, where it was the barbarians, not Romans, who called the tune. The Romans tried to divide and rule with the Huns and Goths, using one as an ally to fight the other and vice versa. To cement these *ad hoc* alliances it was customary to offer high-born hostages as surety against treachery. As part of one of these deals the 12-year-old Aetius was sent as a hostage to Alaric, the formidable Visigoth leader who sacked Rome in 410. Soon after this Rome was in alliance with the Huns and Aetius was shunted on again, this time to the Hun encampment.

Highly intelligent, Aetius used his time with the Huns well, learning their secrets, penetrating their folk-ways, finding the chinks in their military armour. Legend said that he acted as an elder brother to the young Attila and together they hunted in the woods of Hungary. Aetius first occupied the centre of the political stage when the western emperor Honorius died in 423. In the political vacuum when pretenders fought each other for the imperial throne, Aetius backed the wrong horse. His candidate, John, seems to have offered battle prematurely, before Aetius could join him, was defeated by his rival Valentinian and executed. Aetius now came into contact with the second big personality in his career, Honorius's sister Galla

Placidia, who was determined to install her son Valentinian as emperor in the west. Aetius had made elaborate plans to win the throne for John and arrived in Italy with a huge army of Huns, although the figure of 60,000 cited in the ancient sources is a gross exaggeration. Galla Placidia had no wish to engage an army of Huns, so 'pardoned' Aetius, on condition he retired to his estates in Gaul. Valentinian remained as the figurehead emperor but the real power in the west in the years 425–50 was Galla Placidia herself. Recognizing Aetius's talent, she suggested a division of powers, whereby she in effect took all the political and administrative decisions while Aetius took the military ones.

THE FOUNDATIONS FOR ATTILA'S AUTOCRACY

We have no anecdotes or other evidence for Attila's career before 434, but it is a reasonable inference that he honed himself into a mighty warrior, skilled in arms. We can to an extent get inside his mind by reading back from later events to his likely mental processes during the formative years. Under Octar and Rua significant changes occurred that laid the foundations for Attila's later autocracy. In the first place, the multiplicity of Hun chieftains gave way to the idea of a ruler or pair of rulers at the head of the confederacy. Attila noted how fragile this system was, and how the remote Hun chiefs of the Caspian could be subverted by Roman gold, so that any purposive move by the Huns from their new base in Hungary could be countered by the Romans' fomenting a revolt in the east. Some solution had to be found for this perennial problem. Secondly, it was clear that the Huns were no longer so dependent on flocks and the days when they exchanged furs, meat and animals in trade had given way to a new 'monetary' era where the Huns received cash for giving military aid and serving as mercenaries. Thirdly, it was clear to Attila that the Hun empire, stretching as it did from the Caucasus to Denmark was too big to control effectively; the perennial shortage of Hun numbers meant that the remote areas, those most likely to revolt, could not be garrisoned by loyal Huns but by allied troops – the very ones most likely to rebel. Food, too, was a problem, which Rua and Octar tried to solve by abducting peasants and agricultural-ists from their native lands and putting them to work in the new Hun heartland north of the Danube. Something more permanent was needed in this area too. Finally, Attila was aware of the limita-tions of Hungary as the Hun base. The Hungarian plain (some 42,400 square kilometres/16,400 square miles) provided less than four per cent of the grazing land available in Mongolia. Notionally it supplied grazing for 320,000 horses but, to allow room for other animals, the realistic figure was 150,000. If each warrior required ten horses for a campaign – and he did – this meant that Hungary could support only 10,000 warriors.

It was the genius of Attila to go beyond Octar and Rua and real-ize that hiring Huns out as mercenaries was a foolish short-term game. In order to provide the surplus for large numbers of warriors, the previously nomadic horsemen would have to become a perma-nent enforcement agency, which could not happen out on the steppe, with its subsistence economy. But, poised between the west-ern and eastern empires, what was to prevent the Huns demanding money with menaces from both halves of the empire? If Attila could operate a large-scale 'shakedown', he would have the funds to build the Hunnic domain up into a superpower. He could also buy off all opposition from rival Hun chieftains, and co-opt them and his most powerful followers in Hungary with a prize of gold. Uninterested in money himself, Attila knew its power to bind men to a ruler. He could construct a kind of feudalism, whereby subsidiary chiefs were paid handsomely to bring their levies in for a large-scale campaign, and were given large fiefs besides. Octar and Rua had stopped at merely making money out of hiring out military labour. He would make it out of a gigantic protection racket.

Attila is certainly the first clear manifestation in history of the Mafia mentality. As he saw it, the Romans had no choice but to pay him off. It was just too difficult for them to mount effective military operations against the Huns. There were several reasons for this. Because the Huns were so mobile, they could never be caught and could simply move out of range of an enemy host. Moreover, there was no incentive for the Romans to make an all-out effort. Huns were useless as slaves, having no agricultural knowledge and no trade value – which was why Romans tended to kill any Hun prisoners they took. Most of all, the cost of sending an expedition against the Huns was prohibitive; there was no loot or plunder with which to offset the military outlay, as nomads by definition do not live in cities or have fixed wealth and possessions that can be seized. Attila calculated that he could get the Romans to pay protection money provided he was not too grasping. The key was to spread the terror of the Huns' name by selective and well-publicized atrocities so that the enemy would give in without fighting. Attila knew the Huns' manpower problems and realized he simply could not afford to fight too many pitched battles; win or lose, he would waste too many warriors.

ATTILA'S PRECEDENT AS 'MASTER OF SOLDIERS'

Attila came to power in the mid-430s, but to make sense of the events of that decade we have to return briefly to Aetius. He wished to be the West's 'master of soldiers' (supreme commander) but at first Galla Placidia favoured a general named Felix, who was assassinated in obscure circumstances in 428. There was now just one man who could block Aetius's rise to supreme power in the west, Count

Boniface, governor of the important province of North Africa, whom Galla Placidia, never entirely at ease with Aetius, clearly favoured. Gaining the high ground by bringing his army from Gaul into Italy, Aetius persuaded the empress regent that Boniface was planning a coup d'état. He suggested she summon him back to Rome for investigation. At the same time he wrote to Boniface to warn him not to come, as Galla Placidia planned to assassinate him. Clearly Aetius was a machiavellian master.

Worried about his shaky position in North Africa, Boniface made perhaps the greatest single blunder ever made in the entire history of Rome. In Spain the Germanic tribe of Vandals were being hard pressed by the more powerful and numerically superior Visigoths. Their highly talented ruler, Geiseric, had long considered emigrating out of harm's way with his entire people. Suddenly there came an invitation from Boniface to relocate to North Africa as his ally. In a stupendous feat of achievement Geiseric transported all 80,000 of his people across the Straits of Gibraltar to North Africa. But he had no intention of playing second fiddle to Boniface, who suggested they divide North Africa between them; Geiseric wanted it all. Too late Boniface realized he had created an early version of Frankenstein's monster. He sent for help to Constantinople, and the rulers there sent a veteran warrior, Aspar. But in 432 Geiseric decisively defeated Aspar and Boniface. Africa became a forlorn cause, lost to the empire simply through Boniface's stupidity. Aspar, Boniface and the defeated Romans departed for Italy. Now realizing that Aetius had tricked her, Galla Placidia appointed Boniface as supreme commander. Aetius marched to fight him but was heavily defeated; unfortunately for Boniface, he died of wounds sustained in the battle.

Aetius fled to his friends the Huns and secured Rua's backing after intriguing to surrender the province of Pannonia Secunda to him. By now Rua was sole king, as Octar died in 432. With the Huns behind him, Aetius was invincible, so Galla Placidia finally, reluctantly, accepted him as the real power in the western empire. Aetius spent the 430s defending the Gallo-Roman realm in central Gaul, where he was faced by three sets of enemies. On either side of the Rhine was the powerful tribe of the Burgundians, 80,000 strong, with their capital at Worms; in the northwest (roughly modern Brittany) was the ragtag kingdom of the Bagaudae, a kind of refuge for all the dispossessed of the empire: bandits, brigands, runaway slaves led by a man named Tibatto, in some ways a latter-day Spartacus; finally, in the southwest, based on Toulouse, were the ultra-powerful Visigoths, who had pushed north from Spain.

Rua allowed Aetius to recruit Huns by the thousands, and, with them as his cutting edge, he scythed through his enemies in Gaul. The Burgundians seemed to imagine they could do to Aetius what Geiseric had done to Boniface in North Africa. In 435 they swept into Belgium but were checked by Aetius. Determined to finish them off once and for all, in 437 he unleashed a powerful army of Hunnish mercenaries on them. The Huns annihilated the Burgundians, who were said to have lost 20,000 of their fighting men in this one year. Aetius then urged them on against the Bagaudae in late 437, but this campaign petered out in stalemate. Finally, Aetius took aim at his most formidable enemy, the Visigoths of the southwest under their king Theoderic. Fierce fighting in 439 saw the Huns defeated; their fighting record against the Visigoths was never good. Effectively, though, in 433–9 Aetius used the Huns to prop up the ailing Gallo-Roman position in Gaul. Rua's friendship with Aetius had trumped the Huns' self-interest, with little to show for it except mercenary pay. It was this aspect of Hunnish policy under his predecessors that so infuriated Attila.

While Aetius made himself by far the dominant power in the western empire, Rua began to threaten Constantinople, knowing that the bulk of its forces had been earmarked for a joint enterprise by both western and eastern empires to reconquer North Africa from the Vandals. Both the eastern and western emperors agreed that the threat from Geiseric was serious, and when he consolidated his North African conquests by taking the last imperial redoubt at Carthage, the sense of urgency intensified. Rua's demands of Constantinople were modest: simply the return of several tribes who had preferred imperial protection and service under the Byzantine (eastern empire) mantle to his own rule. But Rua suddenly died in 440 and Attila took his chance. Although Rua and Octar had a fourth brother (not Attila's father) who was supposed to be the heir apparent to the Hun empire, Attila and his brother Bleda seized power.

ATTILA'S RISE TO POWER

The details of Attila's rise under Rua in the late 430s are not clear, but it seems that Rua allowed the younger man considerable leeway in expanding and consolidating the Hunnic empire, which by now extended from the Alps to the Baltic and the east of the Rhine to the Caspian. Details on Bleda are even more scanty, but he appears to have been a big, bluff, hard-drinking, hedonistic man, where Attila was small, calculating and far-sighted; pleasure against business, as it were. We also hear of differences over foreign policy, with Bleda opposed to Attila's 'forward policy', which would make sense, for a hedonist has no time for permanent campaigning. As the elder brother, Bleda had a slight seniority, though officially the two siblings were joint kings. It is clear that in many ways Attila found Bleda's roistering tiresome, for we learn of his antipathy for the Moorish dwarf Zerco, kept by Bleda as a kind of court jester – or, some would say, a much tortured mascot. Attila was said to have loathed the very sight of Zerco and seen him as the outward sign of Bleda's stupidity and insignificance.

The emperor at Constantinople at this time was Theodosius, a natural appeaser but a treacherous one, who promised things in treaties and then failed to deliver. Attila brooded over Rua's failure to retaliate when Theodosius welshed on a treaty he had signed with the Huns in 435. He was supposed to have paid a tribute in gold and sent back all the refugees Rua demanded. In fact he sent back a handful of unimportant prisoners, retained all the valuable ones and paid not an ounce of the gold. Attila was determined that things would be different when he was king. The year of his accession, 440, proved a sensational one, with stirring events taking place from the Caspian to the Straits of Gibraltar. In that year Rome and Constantinople collaborated to assemble a huge force, 100,000 strong, for the reconquest of North Africa from the Vandals. Even as this army made its final preparations for embarkation from Sicily, word came in that the Persians had invaded Armenia, in the extreme east of the Byzantine empire. At first a question mark hung over the Sicilian expedition, for the Romans could scarcely fight on two fronts, but then help came from a most unexpected quarter. The so-called White Huns, those who had originally defeated the Alans, were independent of Attila and lived around the Aral Sea, launched an attack on the Persian empire, which compelled the Persians to evacuate Armenia at great speed. Unbelievably, the attack by the White Huns was a spontaneous incursion, not a diversion bought by Byzantine gold. Thinking themselves fortuitously off the hook, the commanders of the African expedition resumed their last-minute preparations in Sicily.

Itching to make his mark and show the eastern empire that a new era had dawned, Attila was suddenly handed an almost miraculous pretext for going to war. He and his brother met the representatives of Constantinople for talks outside the city of Margus on the Danube (in upper Moesia). Famously the Huns held all their meetings while still mounted on their steeds, so the brothers negotiated on horseback; to save face the Byzantines did likewise. In the treaty of 435

Theodosius had promised to pay Rua 350 1b of gold annually but never in fact paid a penny. Attila made it plain that times were changing by doubling the 'tax' to 700 1b and warning that the slightest failure to pay would mean fire and sword. He also stressed that the issue of the refugees had to be settled quickly and that he would brook no compromise on this issue; finally, Roman markets were to be opened to the Huns. Attila wanted to press for even stiffer terms but Bleda overruled him. It was at this point that Attila had his stroke of luck. The bishop of Margus, one of those worldly prelates who put gold first and god a poor second, crossed into the Huns' territory and desecrated some Hunnish tombs in his insane quest for the yellow stuff. In retaliation, in the winter of 440–1, the Huns, admitted to a Roman market, suddenly seized their weapons, killed the guards and all the Roman merchants and seized the nearby fort. Attila and Bleda then announced that unless all refugees, plus the bishop of Margus, were not handed over immediately, this would constitute a declaration of war by the Romans. The bishop of Margus, rightly fearing that Constantinople would sacrifice him to avoid war with the Huns, struck his own deal with Attila: in return for his life and safety he would hand over his city to Attila. This was accepted, and Margus duly fell, but Attila was now on the warpath, determined to press his point about the refugees.

HUN PROWESS IN BATTLE AND SIEGECRAFT

Waiting for the campaigning season proper to begin, Attila swept across the Danube bent on a great war of conquest. A clutch of Roman forts and cities fell rapidly, including the important Roman base at Viminiacum (modern Kostolac). Constantia, Singidunum (Belgrade) and Sirmium were other important centres to come under the Hunnish yoke by the end of 441. The Huns' greatest triumph

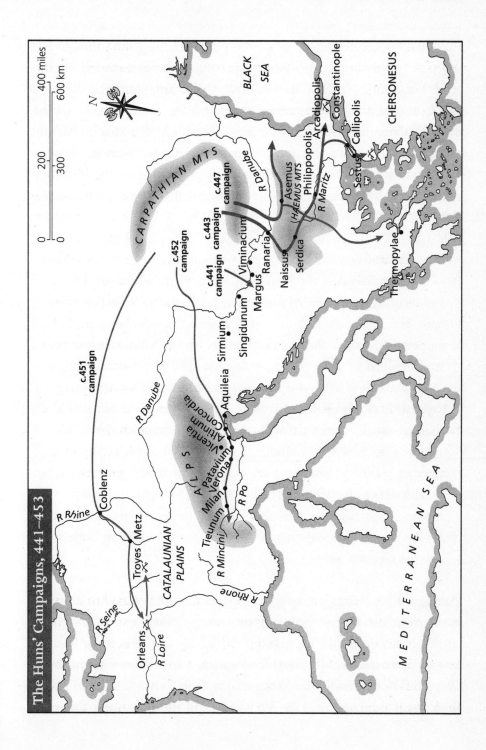

The Huns' Campaigns, 441–453

400 miles
600 km
200
300
0
0

N

BLACK SEA

CARPATHIAN MTS

R Danube

Constantinople

Arcadiopolis

CHERSONESUS

Callipolis

Asemus
(HAEMUS MTS)
Philippopolis

R Maritz

Sestus

c.447 campaign

Naissus
Serdica

Thermopylae

c.443 campaign

Viminacium
Ranaria

c.452 campaign

c.441 campaign

Margus

Singidunum

Sirmium

c.451 campaign

R Danube

Aquileia
Concordia
Altinum
Vicentia
Patavium
Verona
Milan

ALPS

R Po

Tieunum

R Mincini

Coblenz

R Rhine

Metz

Troyes

CATALAUNIAN PLAINS

R Seine

Orleans

R Loire

R Rhone

MEDITERRANEAN SEA

81

was their successful siege of Naissus (modern Nis) early the following year, at the most important crossroads in the Balkans, where one
road led south to Thessalonica and the other southeast to Sofia
and Constantinople. This proved that the Huns were by now formidable not just in battle but also in siegecraft, and that they had the
capacity to take the strongest Roman fortresses, such as those at
Viminacium and Naissus. The Roman historian Priscus, the indispensable source for the career of Attila, described the siege of Naissus as
follows:

> When a large number of siege engines had been brought up to the
> wall ... the defenders on the battlements gave in because of the
> clouds of missiles and evacuated their positions; the so-called rams
> were brought up also. This is a very large machine. A beam is
> suspended by slack chains from timbers which incline together, and
> it is provided by a sharp metal point and screens ... for the safety of
> those working it. With short ropes attached to the rear, men
> vigorously swing the beam away from the target of the blow and
> then release it ... From the walls the defenders tumbled down
> wagon-sized boulders ... Some [rams] they crushed together with
> the men working them, but they could not hold out against the
> great number of machines. Then the enemy brought up scaling
> ladders ... The barbarians entered through the part of the circuit wall
> broken by the blows of the rams and also over the scaling ladders ...
> and the city was taken.

The new-found Hun prowess in siegecraft was an alarming development from the Roman point of view and marked a significant shift
in the balance of military power. It had always been assumed that
'mere' barbarians might be able to spring a treacherous ambush but
they could not besiege and take major cities; after all, the Goths in
their great rampage in 376–82 had found the Byzantine fortresses

beyond them. How, then, had Attila made such an advance? Some say the Huns already possessed the secrets of siegecraft when they first burst into Europe from the Asiatic steppes, but this is highly unlikely. More likely, this is a sea change that can be attributed to Attila himself. His men had fought as auxiliaries when Aetius and the Romans besieged the Visigoths in Toulouse in 439 and they had studied Roman tactics at first hand, learning the new skills as ordered by Attila. That the Huns under Attila had moved up a military notch from the days of Rua was obvious from the skilful way they went about securing the capitulation of Viminacium and Naissus, but it was the innovations of Rua and Octar that had in another sense made this success possible: it was only their mass use of forcibly migrated labour in the fields that released new volumes of fighting men for the armies. As is well known, siegecraft operations are labour-intensive, requiring huge amounts of manpower for digging trenches, sapping, manning the siege engines, and so on. Attila showed he meant business in his tough demands by having a credible army to back up his threats. The point was not lost on Byzantium that the balmy days of Rua were over; if they did not pay this new king, they would have his sophisticated new army with which to deal.

ATTILA'S BALKAN CAMPAIGN

The effect of Attila's whirlwind Balkan campaign of 441–2 was immediate. Geiseric the Vandal king got a sudden reprieve as a panic-stricken Constantinople recalled its forces from Sicily. But these troops did not get back in time to save the Byzantines from having to make a humiliating peace with Attila. He showed how he meant to go on by immediately doubling the protection money he required from 700 lb to 1400 lb of gold. He also demanded 7000 lb of gold as arrears and insisted that every last refugee from the Hunnic domain be handed over on pain of a renewal of hostilities. But the inevitable happened. As soon as their crack troops returned from Sicily, the Byzantines stopped paying the tribute, instead using the money to recruit a large force of Isaurian mercenaries from Cilicia in southeastern Asia Minor. Attila went to war again in 443 and again took a raft of cities in the Balkans. Encountering Constantinople's best generals in the Chersonesus, he defeated them all: Aspar, Areobinesus, Arnegisgus. While continuing to demand the entire arrears of the gold payments, Attila sent constant embassies to Constantinople, nagging about the refugee problem. By this time he realized that Byzantine custom dictated lavish gifts for ambassadors, to be taken home to their sovereigns. By sending dozens of envoys on every conceivable pretext, including a demand for the return of 'most wanted' individuals who were no longer in the territory of the eastern empire, Attila ensured a constant flow of wealth back to his base in Hungary. With the eastern emperor and his advisers in despair, the intermittent nature of the Hun threat once more manifested itself. In 445 Attila had his brother Bleda murdered as part of a wider power struggle, reflecting divided counsels among the Huns. For two years Attila was too busy asserting sole authority over his far-flung empire to deal with the Byzantines. Their hopes rose in turn, enhanced by rumours of serious rifts in the Hunnic high command.

Theodosius used the breathing space to make Constantinople impregnable. In 413 the new Theodosian Walls were constructed to run from river to sea, to cover the entire area from the Sea of Marmara to the Golden Horn inlet. It took nine years to build the triple line of walls and now, in 445–7, the emperor was determined to plug any gaps in the defence and strengthen the defences even more. Some 5 kilometres (3 miles) long, the walls rose tier by tier, with stones at the base and then mounting like a stairway, to form a barricade 60 metres (200 feet) thick and over 30 metres (100 feet) high. The first obstacle to any attacker was a 20-metre (60-foot) wide moat, 10 metres (30 feet) deep, partitioned by locks, which allowed it to be flooded. Then came a 20-metre (60-foot) wide parapet lined with defenders. Next was the outer wall, 10 metres (30 feet) high, with a walkway on top patrolled by guards. Beyond this stretched a 15-metre (45-foot) wide parapet and then a 20-metre (60-foot) inner wall, again with a walkway and with towers placed at 45-metre (150-foot) intervals. Each of the ten city gates had a drawbridge that was withdrawn entirely during a siege.

Not even the Huns could hope to overcome such elaborate defences and, as fortune would have it, their one chance for a breach in the walls occurred while they were far away. In January 447 Attila was at last ready to resume hostility with the Byzantines but, before he could advance on Constantinople, nature took a hand. On 27 January 447, between 1 a.m. and 2 a.m., a violent earthquake struck Constantinople, laying the area around the Golden Horn in ruins; an entire section of the great walls collapsed. The Byzantines recovered with amazing rapidity. Constantius, the Praetorian Prefect of the Guard, mobilized the circus factions (the ancient world's equivalent of football supporters) to put enmities aside and collaborate in the rebuilding of gates and towers and clearing the moats of rubble. By the end of March all the damage had been repaired; Constantinople was itself again long before Attila could arrive outside its walls.

THE WRATH OF ATTILA IN THE EAST

Yet if the eastern capital had had a lucky escape, the rest of the empire felt the wrath of Attila. The year 447 was a terrible one for Roman arms and Roman civilians. Attila gave Theodosius a chance to avoid bloodshed by meeting the financial terms at once and in full, but the emperor jibbed at the level of taxation he would have to impose. Some historians have argued that Attila's financial demands, though hefty, were not as ruinous as the Byzantines claimed, especially if the alternative was raising taxes to fight prolonged wars. Attila, as we have seen, demanded 7000 lb of gold in arrears and a subsidy of 1400 lb a year. This approximates to what one might call the market rate for such levies. Alaric, the Visigoth leader, was paid off by Rome in 408 with 9000 lb of gold, while the Persians in 540–61 received 1000 lb a year from Constantinople. Given that the annual revenue of the eastern empire in the late 440s was of the order of 270,000 lb of gold, Attila was asking for about 2.2 per cent of the Gross National Product as the arrears payment and about 1 per cent a year thereafter.

But, as always with blackmail, shakedown and protection money, it is the principle that so irks the donor. This time even Theodosius the appeaser thought that enough was enough; he told Attila's envoys he would negotiate but that was all and there would be no giving in to threats. To deflect Attila's attention the Byzantines stirred up trouble in his rear, enlisting the tribe called the Acatziri (who lived either on the shores of the Baltic or on the Sea of Azov; unfortunately the sources are not clear). Unhappily for the Byzantines, Attila dealt with this backdoor threat easily. The Acatziri learnt to their cost, as had so many of the tribes the Romans enticed to rebel against the Huns, that their tempters could do nothing for them when the Huns decided to take revenge.

The war that followed was a disaster for the eastern empire. The Huns destroyed frontier forts along the Danube, and took Ratiaria, then advanced along the river to the north of the Haemus mountains. Meanwhile the Roman general Arnesiscus, commander of the imperial forces, advanced northeast from his headquarters at Marcianople and drew up his forces at the River Utus (modern Vit) in Romania. The battle that followed was hard fought and close run.

This time Attila was faced not by the levies he had conquered so ignominiously in the earlier campaigns but with the cream of the Byzantine army. Finally the Romans were overwhelmed but the Hun victory had been pyrrhic: they had taken heavy casualties, and some historians believe the Hun military machine never recovered from the manpower losses at the Utus. But in the short term victory opened the mountain passes; the Huns swarmed south on to the plains of Thrace. There was a second battle of Chersonesus, which the Huns won.

The roll-call of cities that fell to Attila included virtually everywhere of significance in the European part of the Byzantine empire, except Adrianople and Heracleia, and most of them were razed, utterly destroyed: Marcianople, Arcadiopolis, Philippopolis (modern Plovdiv), Constantia, Nicopolis. When Attila's armies reached the Dardanelles and the coast of the Black Sea, bivouacking at Sestos and Gallipoli, Attila could confidently state that he now controlled the entire Balkans. His men had ridden 100 kilometres (600 miles) from the Hungarian plain to the outskirts of Constantinople. St Hypatius of Bulgaria said the Huns captured a hundred cities and so terrified the eastern empire that even the monks at Constantinople wanted to flee to Jerusalem. There can be no doubting the devastation. 'They [the Huns] so devastated Thrace,' Hypatius said, 'that it will never rise again and be as it was before.' Three hundred years of the Romanized Balkans were destroyed almost overnight. And this was not like the cosy sack of Rome in 410 by a fellow-Christian Alaric, whose troops did little damage to either persons or property. Nicopolis, one of the showpieces of the Roman Balkans, was wiped out, so that not a stone remained on a stone. This was true wrath-of-God stuff. The town of Asemus, perched high on an impregnable hilltop, was one of the few to survive the whirlwind.

Beaten and humiliated, the Byzantines were forced to sue for peace. The years 448–9 were taken up with protracted negotiations, with the Byzantines on the back foot while the Hun leader sat back and enjoyed himself. Attila now demanded a subsidy of 2100 lb of gold a year and produced a fresh list of wanted 'criminals' that Constantinople was said to be harbouring; this of course would mean more embassies and more lavish gifts. Attila now for the first time really began to put the bite on the Byzantine élite, for to raise such sums (not to mention the 7000 lb of gold in arrears that was still outstanding), the landowning oligarchy finally had to waive their untaxed status. Despairing of a military solution and possibly now themselves under threat from the landowning class they had represented but were now forced to sacrifice, Theodosius and his eunuch prime minister Chrysaphius came up with the oldest expedient of all: an assassination plot. Attila's chief envoy to Constantinople and one of his inner circle of élite advisers, the so-called *logades*, was a man named Edeco, and he it was whom Chrysaphius now tried to turn.

A MURDER PLOT AGAINST ATTILA

Edeco agreed to murder Attila and set out for Hungary in company with an embassy under the Byzantine diplomat Maximinius, which would act as a blind to divert Attila's attention while Edeco finalized the murder plot. The unfortunate Maximinius was left out on a limb, being told nothing about Chrysaphius's real designs. He and his party took 17 of the demanded refugees and sackloads of presents for Attila – silks, pearls, gold and personal gifts – and he himself was given instructions to negotiate toughly with Attila over the Hun leader's latest demand: permanent annexation of all lands adjacent to the Danube and a demilitarized zone to prevent Byzantine counter-attacks. In many ways Maximinius was involved in a low farce – a phoney embassy that was never meant to succeed. But his journey to Attila's camp and his adventures there are of supreme importance to the historian. As his assistant, Maximinius took the historian Priscus, who provides us with an incomparable picture of the Huns' everyday life as well as the only authentic close-up view of Attila himself.

The Roman envoys toiled in a northwesterly direction from Constantinople, passing through Serdica, the Succi pass and Naissus, where the bones of those slaughtered in 441–2 could still be seen bleaching on the ground outside the devastated city. They handed over five of the 17 wanted men, who were put to death by impalement – having a stake driven up through the anus. After striking due north through woods and wastelands, they came to the Danube, were ferried across by Hunnic boatmen and then escorted 16 kilometres (10 miles) to Attila's camp. Maximinius and Priscus had intended to target one of the *logades* called Onegesius, supposedly a 'dove' among the hawks on Attila's council, but now learnt that Onegesius was himself absent on a mission.

A still greater shock awaited them when Scottas, one of Attila's right-hand men, ordered them to leave at once as there was nothing to discuss; Attila knew all their proposals, and there was nothing in them that interested him. The quick-thinking Priscus offered Scottas a bribe to get them in to see Attila, but they were still puzzled how Attila knew every last detail of their mission. It later transpired that Edeco had gone straight to Attila and revealed the assassination plot and all the details of the official embassy. One of the key elements in the conspiracy was that the official Roman interpreter Vigilas (or Bigilas) would return to Constantinople, once Edeco had hatched the plot, and secretly bring back 50 lb in gold to bribe Attila's guards. When Scottas secured the desired interview for Maximinius and his party, the first thing that happened was that Attila raged viciously against Vigilas. This seemed to Maximinius and Priscus like typical barbarian irrationality, but of course they did not know what Attila knew. Attila then turned to Maximinius and told him coldly that no proposals from Byzantine embassies would ever be entertained until every last political refugee was returned to him.

A MAN CONFIDENT OF HIS OWN DESTINY

Despite his public tantrums, Attila allowed the Romans to trail around behind him for two weeks, and it was in this period that Priscus recorded his priceless observations about the Huns and Attila himself. The Huns lived in walled compounds like very large villages. As Priscus noted: 'Inside the walls there was a large cluster of buildings, some made of planks and fitted together for ornamental effect, others from timbers which had had the bark stripped off and planed straight. They were set on circular piles made of stones, which began from the ground and rose to a moderate height.'

Attila lived in the largest dwelling, which was embellished with towers to distinguish it from the others. The living quarters were carpeted with woollen–felt rugs, and the Hun women worked coloured embroidery and fine linens. The Huns dined from couches in the Roman style and had tapestries and multicoloured ornamental hangings to decorate their rooms. Although the Hun warriors lived polygamously, women are thought to have had a higher status among the Huns than in most nomadic, steppe or Oriental societies. Advocates of polygamy, like the nineteenth-century explorer Sir Richard Burton, liked to say that in primitive societies women's lot was actually improved by polygamy. The proposition may be contentious, but it is the case that the wife of Bleda, Attila's murdered brother, was living a highly privileged and respected life in Attila's encampment; to kill a woman's husband clearly did not imply disgrace or ostracism of the wife among the Huns. Bleda's wife (whose name has not been recorded) actually impressed the Roman envoys

with her femininity and kindness, on one occasion allowing them to
dry out in her quarters after a downpour.

Nevertheless, in a warrior society the male ethos was paramount,
and some of Priscus's descriptions show women playing no more
than a 'cheerleader' role. Attila's entry into his dining hall is
described thus:

> As Attila was entering, young girls came to meet him and went
> before him in rows under narrow cloths of white linen which were
> held up by the hands of the women on either side. These cloths
> were stretched out to such a length that under each one seven or
> more girls walked. There were many such rows of women under
> the cloths and they sang Scythian songs.

Attila himself did not impress Priscus so much at first sight. Short of
stature, broadchested, with a huge face and squat body, he had small,
deep-set eyes, a flat nose and a wispy beard flecked with grey. He
strode around the hall with an arrogant gait, glancing insolently
from one side to the other. Mercurial, moody, suspicious, Attila
seemed both abstemious with wine and humourless: jesters made
the other Huns laugh but Attila sat there, brooding, sullen and
unsmiling. The only time Attila showed a human side was when his
youngest son, Ernac, the beloved Benjamin of his family, entered the
banqueting hall. Priscus noticed Attila's hard eyes soften immediately; he called the youth to him and stroked his cheek. Why was he
the favourite, Priscus asked? It seemed that a soothsayer had once
told Attila that his family would decline after his death, but that
Ernac would restore the fortunes of the Huns. Deeply superstitious,
addicted to auspices, oracles, shamans and medicine men, Attila took
this very seriously.

Yet, despite himself, the more Priscus saw of Attila, the more he
was impressed. He was not just a brutal thug and cruel tyrant, but

evinced shrewdness and even wisdom: 'Though a lover of war, he was not prone to violence. He was a very wise counsellor, merciful to those who sought it and loyal to those he had accepted as friends.' He had no use for wealth except as a means of political and social control, and his own tastes ran to austerity rather than conspicuous consumption. Here is Priscus again:

> While for the other barbarians and for us there were lavishly prepared dishes served on silver platters, for Attila there was only meat on a wooden plate ... Gold and silver goblets were handed to the men at the feast, whereas his cup was of wood. His clothing was plain and differed not at all from the rest, except that it was clean. Neither the sword that hung at his side nor the fastenings of his barbarian boots nor his horse's bridle was adorned, like those of the other Scythians, with gold or precious stones.

Attila was thus sending out a subtle message. He did not need the outward signs and trappings of power, since he was so confident of his own destiny. He was also telling his supporters that he could never be bought with Roman gold, and that austerity alone bred great warriors. Attila was ruthless and would dispatch known enemies in the twinkling of an eye, but he was also intelligent and realized that mere brutality eventually alienated even close disciples: he knew enough of the history of Rome to remember the emperors who had perished because of their mindless cruelty: Caligula, Nero, Domitian, Commodus, Elagabulus. His followers had to know that he was tough, and he gave them many proofs of that; but they also had to realize that he had a capacity for moderation, compromise and respect for his subordinates that would make it worthwhile for them to enter into lifetime partnership with him.

ATTILA THE PERFECT AUTOCRAT

Attila could feel confident by the late 440s as he had emancipated himself from all traditional tribal obligations and limitations; he had even killed his own brother, against all the mores and folk-ways of the Huns. He was worshipped like a god and had no limitations on his power. No councils, assemblies or senates could constrain him, he could pass final judgement in person on anyone on the spur of the moment, he had the power of life and death over all Huns; in a word, he was the perfect autocrat. All this he had achieved by extracting a tithe from the Byzantines and binding his followers to him with threads of gold. The effect of this was to transform what had once been a nomadic society into a rough-and-ready kind of feudalism, where his *logades*, men like Onegesius (his Number Two man), Berichus (Number Three), Oebarsius (Attila's paternal uncle), Orestes, Edeco and Scottas, had 'fiefs', and they were Attila's satraps or governors. Attila kept them on their toes by giving them territories unequal in area, population, wealth and strategic importance to hold and encouraging them to compete in the collection of tribute and food from the subject territories. Attila also shrewdly obliged the subject territories to send him all their men of military age when he went on campaign, so that there could be no 'stabs in the back' or rebellions in his rear; manpower shortage meant the Huns could never campaign and garrison the conquered territories. Attila promoted and pampered key allies, such as Ardaric, king of the Gepids, and Valamar of the Ostrogoths, so that they would never be tempted to subvert his absolute power.

Attila also had the important attribute of patience and was prepared to bide his time for maximum effect, awaiting the precise moment for a piece of theatre. He could have arrested and even executed Vigilas for the assassination plot Edeco had revealed to him. Instead he sent him back to Constantinople, ostensibly to get further details of the proposed peace treaty but really to gull Vigilas to bring back the blood money and thus catch him red-handed. To make sure that there could be no slithering out of the trap, he expressly ordered that no one entering Hunnish territory could bring gold or other precious metals with him. Maximinius and Priscus set out on their return journey and ran into Vigilas on his way back from Constantinople with the loot. When Vigilas returned with the 50 1b of gold, thinking the plot was still in being, Attila's men searched his baggage and found the stash. He was haled before Attila, who demanded an explanation. Vigilas started to bluster about his status as an ambassador but had made the bad mistake of bringing back his 20-year-old son with him. Attila sprang up, held a sword to the boy's throat and said he would kill him unless Vigilas told the whole truth. The wretched Vigilas, distraught and in tears, made a full confession.

Attila held him and his son as hostages pending 'satisfaction' from Constantinople. He then sent Orestes to the Byzantine capital to seek audience with Theodosius and Chrysaphius. In the emperor's presence, Orestes produced the bag that had originally held the gold (Attila did not send that back) and asked the emperor if he recognized it. He then repeated a verbatim message from Attila. In Priscus's words:

> Theodosius was the son of a nobly born father, and Attila too was of
> noble descent ... But whereas Attila had preserved his noble lineage,
> Theodosius had fallen from his and was Attila's slave, bound to the
> payment of tribute. Therefore, in attacking him covertly like a
> worthless slave, he was acting unjustly towards his better, whom
> fortune had made his master.

Thoroughly enjoying himself, Attila next sent to Constantinople to demand that his would-be murderer Chrysaphius be delivered up to him. He knew the demand would be refused, but that gave him the excuse to send more envoys to the Byzantines, who would have to be 'honoured' with gold, which would make its way back to Hun headquarters in Hungary. To placate him, early in 450, Chrysaphius sent him two ambassadors who were known to be at the top-level of Byzantine diplomacy, only ever sent on the most important missions. The two men were Anatolius and Nomus, whom Attila had explicitly requested by name. Perhaps more to the point, Chrysaphius sent a huge quantity of gold, asking Attila to lift the death sentence he had placed on him.

To everyone's surprise, the embassy was a huge success. Attila actually honoured the envoys by crossing the Danube to meet them. All outstanding issues were resolved. Attila promised to launch no more attacks on the eastern empire, to withdraw from the area immediately around the Danube and to rescind his previous demand for a demilitarized zone around that river. He even released Vigilas and his son to show his goodwill. Everyone was astonished. What was in the offing? What was Attila planning? When he was in Attila's camp, Priscus had heard wild talk of a Hun invasion of the Persian empire and had thought it too good to be true, for such an endeavour would surely cripple Hunnic power once and for all and play into the Romans' hands. Yet even as the bemused Byzantines wondered what would be next and could scarcely believe their own good luck, the wheel of fortune turned once more. In July 450 Theodosius fell from his horse and died from his injuries. The new emperor quickly emerged as one Marcian, a tough soldier and veteran of many campaigns, including the failed one by Aspar against Geiseric in North Africa. Marcian immediately announced that the era of blackmail and extortion was over. There would be no more gold payments to Attila; if he wanted peace, well and good; but if not, Constantinople was prepared to fight to the death.

A WAR OF ATTRITION

Marcian had a resolute quality that belied his 60 years. The famous historian Edward Gibbon described him thus: 'It was the opinion of Marcian that war should be avoided as long as it is possible to preserve and secure an honourable peace; but it was likewise his opinion that peace cannot be honourable or secure, if the sovereign betrays a pusillanimous aversion to war.' In other words, Marcian despised his predecessor Theodosius's appeasement policies. He could see right through Attila and his policy of demanding money with menaces but, as he saw it, Constantinople had powerful cards that it had not played. The way to deal with Attila was not to offer pitched battles on terrain where his mounted archers had the advantage nor to roll over and give him whatever he demanded. The key was to wage a war of attrition, for the Huns could not sustain constant manpower losses in battle. The resources of the Byzantine empire made it better placed to succeed in a long, drawn-out struggle.

Without gold, Attila could not keep his key chiefs and retainers sweet, and once factionalism arose within the Hun ruling circles, Constantinople could no longer be overawed or dominated. Marcian therefore tightened the screws as much as he could. In addition to stopping the flow of gold to the Huns, he closed all the imperial markets to them, in effect beginning a slow economic blockade to destroy their wealth. He forbade the export of arms to the Huns by anyone within the empire, knowing that Attila could not make enough of his own and that the Huns' way of making war was 'arrow-intensive'. Finally, he sent his secret agents out to bribe the leaders of the Gepids and the Ostrogoths and to encourage revolts against Attila. We should mention in passing that Marcian's advent meant the triumph of the landowning classes in the Byzantine empire. The pro-Hun appeasement policy of Theodosius had been backed by merchants, traders and manufacturers, who could make money out of the barbarians, but vehemently opposed by the landowners who had to stump up to pay Attila's gold. Marcian's change of policy was not just a personal thing but represented the reassertion of power by the empire's landowners.

The year 450 saw Attila inactive. It is clear that he was seriously disconcerted by the new wind blowing from Constantinople and grasped immediately the nature of the threat from Marcian. But he also knew that he was not strong enough on his own to humble Constantinople. What to do? This is the context in which we have to understand the (at first) inexplicable decision to make war in the west and march to Gaul in 451. The charm offensive with Anatolius and Nomus and the surprising decision to make peace with Constantinople may even reflect a decision Attila had taken before Theodosius died. Since he, Attila, was not strong enough to defeat the Byzantines singlehanded, he had to look around for possible partners. He thought back to 442 when his offensive had led the Romans to call off their joint expedition against the Vandals in North Africa. It seemed clear that the one thing that could paralyse both the western and eastern empires was collaboration between Huns and Vandals.

Since the Huns were not a literate society, we cannot go to archives to find evidence of diplomatic parleys between Attila and Geiseric in 450–1, but we cannot make sense of the events of 451 without assuming them. Besides, the historian Johannes tells us that there were such contacts. Motive, means and opportunity all point in the same direction. The one enemy Geiseric really feared was the Visigoths of Spain; after all that was why he had initially migrated across the Straits of Gibraltar. But he was not strong enough on his own to deal with them decisively, just as Attila was not strong enough unaided to defeat the Byzantines. But a Vandal–Hun alliance could change the picture completely. By this time Geiseric had built up a powerful navy, which had already defeated the western Romans.

Attila probably envisaged an amphibious attack on Constantinople by Huns and Vandals as his golden dream and put the proposal to Geiseric. The Vandal king, however, the cleverest politician in the Mediterranean at this time, needed a quid pro quo. If Attila were prepared to destroy his enemies the Visigoths in the west, the two of them could then march on Rome, making Geiseric the lord of the western Mediterranean. In return Geiseric would send his powerful fleet to assault Constantinople while Attila attacked from the landward side; total victory would then confirm Attila as lord of the east. In effect Geiseric and Attila would be the two new Roman emperors.

We should expect the master diplomat Geiseric to win a battle of wits with Attila, and so it proved. It is clear that in the course of protracted negotiations Geiseric insisted that Attila had to help him first. Attila agreed and announced that he would be campaigning in 451 against the Visigoths of southwestern France. He realized it was essential that his attack on the Visigoths should not suck in the Romans as well, so sent an embassy to the western emperor Valentinian in Italy to emphasize that his campaign was directed solely against the Visigoths, that he had no quarrel with Rome. Political analysts at the time must have been able to work out that an attack by Attila on the Visigoths could benefit Geiseric only, but they were blindsided because Geiseric currently enjoyed friendly relations with the western Romans he had thrown out of North Africa.

It was of course known that the Visigoths and Vandals were mortal enemies. Quite apart from the geopolitical struggle for mastery in the extreme western Mediterranean, there was bad blood between Geiseric and the Visigoth king Theoderic. Geiseric's son Huneric married Theoderic's daughter and turned out to be a drunken wife-beater. In one of his rages he destroyed her beauty by cutting off her nose and sent her back to Theoderic. The Visigoth king vowed revenge but had his hands full meanwhile with Aetius and his Gallo-Roman supporters in central France. Huneric may additionally have been a reason for a subtle cooling in relations between Geiseric and the Roman emperor Valentinian. As part of a wider peace settlement in the mid-440s, it was agreed that Huneric, now conveniently free of his Visigoth bride, should marry Valentinian's daughter Eudocia when she came of age. But when Eudocia reached her twelfth birthday, the Romans showed no signs of honouring the deal. First Valentinian said he had had second thoughts and decided to substitute his younger daughter Placidia as the sacrificial bride. Then, under pressure from Aetius, still the real power in the western empire, he announced that Placidia would be married to Aetius's son Gaudentius. Construing this as a slap in the face, Geiseric was in a broodingly anti-Roman mood when Attila's envoys came to call.

Even with Geiseric's support, Attila's decision to march west was risky. The Huns basically lacked the planning, administration and logistics for long campaigns and everything was predicated on an enemy being terrified by their very name; they had not thought through what to do if the enemy refused to roll over before them. Nor had Attila faced up to the fact that western Europe was a very different physical habitat and environment from the one he enjoyed on the Hungarian plain. The last great grassland in a westerly direction could support large numbers of horses, but Gaul could not. It was the lack of forage and pasture that would turn the Mongols back from a conquest of Europe in 1242 once they had reached Hungary. In short, western Europe was naturally heavily biased in favour of defenders and against nomadic invaders. Attila may also have discounted the physical toll on his troops. In 447 he had made them march over 500 kilometres (300 miles), and it had been a grim business, but now he was proposing a march of over 1200 kilometres (750 miles). On the other hand, without Geiseric's help it made no sense to march on Constantinople. The triple land walls of the city made it impregnable – indeed they would keep it safe until 1453, when the Turks finally blasted a way in. The Byzantines would be fighting any war on interior lines, with ready access to all their tax-rich provinces throughout Turkey, south to Egypt and beyond. Attila needed a navy so that he could land troops on the Turkish mainland of the Byzantine empire and get round behind Constantinople. Geiseric had the answer, but insisted that Attila settle with the Visigoths first.

BREACH WITH AETIUS AND THE WESTERN ROMANS

Even so, on certain projections things looked good for Attila's grand design against the Visigoths, but then two things happened to alter the course of history. Aetius, previously the friend of the Huns and enemy of the Visigoths, turned against Attila. While Attila was terrorizing the eastern empire in the 440s, Aetius was continuing to make war on the Visigoths and the Bagaudae of the northwest who, despite many reverses, seemed to be growing stronger all the time. In 442 he settled a body of Alans near Orléans and encouraged their king to make war on the Bagaudae. The Roman position in Gaul was increasingly precarious. By 450 they still controlled the Mediterranean coast but inland held merely a central strip, bounded on the north and west by the Loire, plus the whole of the Rhone valley. The Burgundians controlled the Alps and Savoy, the Franks what is now northeastern France, and the Bagaudae the northwest as far down as Poitou and Anjou, where the Visigoth kingdom of Aquitaine began. Moreover, by this date the western empire had lost Spain and Britain.

Continuing low-level warfare seems to have marked this decade in the history of France, but by 450 Aetius was still the implacable enemy of the Bagaudae and Visigoths. Throughout this time Aetius remained suspicious of Attila's ultimate intentions, and there are indications that he made contingency plans for a Hun invasion of Gaul as early as 442. It was around 449 that Attila first evinced signs of hostility to his old friend and reliable Hun ally. First, he gave sanctuary to a Bagaudae leader defeated by Aetius in 448. Then he backed a different candidate as king of the Ripuarian Franks – yet another tribe that had settled in Gaul – from the one Aetius nominated. Moreover, Attila had already shown a disturbing tendency to become obsessed with comparatively trivial matters and to build them up to diplomatic crises. There was a tedious and complex affair concerning some gold plate Attila claimed as his property but which a Roman banker named Silvanus had sold and then pocketed the proceeds. Attila took the line that Aetius was the real power in the western empire, so the fact that he had done nothing about the Silvanus affair showed his basic hostility. But the net effect of Attila's behaviour was to make Aetius start to think the unthinkable: perhaps the way forward was to forge an alliance with the Visigoths to counter the growing threat from the Huns.

The second incident once again shows Attila's poor judgement. Just before he set off for the west, Attila received a sensational message from Justa Grata Honoria, sister of Valentinian. Honoria was seduced by a steward named Eugenius, who was executed for lèse-majesté. Valentinian then decided to marry off his troublesome sister to an obscure senator. Unwilling to accept her lot, Honoria wrote to Attila, offering herself as his bride and promising him a hefty sum if he would come and rescue her from her dreadful plight. Valentinian was advised to hand her over to Attila to remove any pretext for an attack on Rome but Aetius was violently opposed, since an Attila–Honoria match would cut the ground from under his feet. So Valentinian instead gave her in charge to her mother Galla Placidia. Honoria then disappeared from history; all the indications are that she was later secretly executed as an enemy of the state. But Attila then claimed that the Romans had kidnapped his 'wife' and angrily demanded her restitution or massive compensation. Moreover, he claimed that by the fact of his 'marriage' to Honoria, he was now the rightful heir to the western Roman empire and should immediately be recognized as 'master of the imperial armies' in place of Aetius.

By November 450 both Aetius and the western Romans were resigned to an outright breach with Attila. By his crazy antics over Honoria, Attila made it likely that the Romans and the Visigoths might seek common cause. Even worse, by claiming the western Roman empire as his own, he seemed to have disowned the prior agreement with Geiseric that he had worked so hard to build up. Geiseric was disillusioned and made no move to attack the Visigoth flank when Attila marched against them. This explains the 'dog that barked in the night' that has so puzzled historians sceptical of an Attila–Geiseric accord. There was a pact to divide the two Roman empires but, by seeming to claim the western empire as his, Attila effectively welshed on the deal.

'I AM THE WRATH OF GOD'

Aetius meanwhile had two main problems. He had to persuade the Visigoths under Theoderic to forget his own enmity towards them for the past 20 years, and he had to persuade the Visigoths to campaign further north in France, both to prevent the Gallo-Roman lands being ravaged and to deny the resources of the rest of Gaul to the Huns. Aetius left Rome in command of a very small force – small because Rome was being ravaged by famine and disease. To secure the alliance with Theoderic he engaged his friend Avitus, scholar, diplomat and future Roman emperor, also, and most importantly, a trusted friend of Theoderic; Avitus took warm messages of friendship from both Aetius and Valentinian. Theoderic did not need much persuasion: as he saw it, if the Visigoths had to go down fighting before Attila, it was best if they did so with the protection of powerful allies. Meanwhile Aetius, in marked contrast to Attila, showed his diplomatic skills by assembling a coalition of all the French tribes to oppose the Huns: the army converging on central France would contain Romans, Visigoths, Ripuarian Franks, Salian Franks, Burgundians, Saxons and the Bagaudae. The odds were beginning to tilt against Attila, but the all-France alliance was not his only problem. He needed to get to Gaul fast, which meant leaving behind his heavy siege engines and catapults. But going fast meant leaving in his rear powerful garrisons who could sever his lines of communication. In any case, the nature of Hunnic warfare meant he could not ride for the French frontier as fast as he wanted: he needed wagons for the hundreds of thousands of arrows his warriors would fire. His advance was therefore a half-and-half affair; he did not attain the speeds he desired but he still left the garrisons in his rear.

In the spring of 451 Attila marched west from Hungary. His army may have been 30,000–40,000 strong – a huge figure for that era. Typically, ancient historians, with wild hyperbole, spoke of half a million men – a number beyond the capacity of the whole of Europe to feed. The contemporary poet Sidonius Apollinaris tried to suggest the polyglot, heterogeneous nature of the Hun army with the forced exactitude of rhymed metre: 'Suddenly the barbarian world, rent by mighty upheaval, poured into the whole north of Gaul. After the warlike Rugian comes the fierce Gepid, with the Gelonian close by; the Burgundian urges on the Scirian; forward rush the Hun, the Bellonotian, the Neurian, the Bastarnian, the Thuringian, the Bructeran and the Frank.'

The truth is that, while one or two members of some of these races might have been present, the Huns, as usual, relied heavily on the Ostrogoths and the Gepids. The immense force followed the upper Danube southwest out of the Hungarian plain, marching along both sides of the river. Near the headwaters the army split into two, with the main body following the passage from the Danube to the Rhine, while a subsidiary detachment went via Basel, Strasbourg, Speyer, Worms and Mainz to overawe the Franks; but the Huns were too late, since most of the Franks had already gone west to join Aetius. The two forces rendezvoused on the Rhine and crossed the river near Coblenz. Traditionally, the route through Luxembourg was the only viable one for large armies, since the Vosges, Ardennes and Eifel mountains protected France against westward crossings further south on the Rhine. Attila now bypassed the fortress of Trier and laid siege to Metz, which fell on 8 April 451. The Huns then covered the 300 kilometres (200 miles) from Metz to Orléans in three weeks, arriving there in May. It was in this period that Attila took Troyes. The traditional story says that bishop Lupus of Troyes went out to ask Attila for mild surrender terms and told him he was a man of God. Attila then replied: '*Ego sum flagellum Dei*' (I am the wrath of God), and this was to become part of the enduring black Christian legend of Attila.

Attila's lightning campaign secured him the initial advantage, for the allies had not yet rendezvoused; Aetius had only got as far north as Arles. Attila now closed in on Orléans, encouraged by a message from the treacherous Alan leader Sangibanus that he would betray the city to the Huns for a consideration. But Sangibanus, hearing conflicting reports of the progress of the allied army, dithered, and Attila did not have the siege engines to finish off Orléans quickly. It was June before the city was ready to surrender, and at this juncture, if we are to believe the legends, miraculous events started to multiply. First, the obscure little town of Paris was saved a visit from the invaders by the prayers of St Geneviève. Then, just as Orléans had surrendered and the first Huns were riding through the gates, sentries on the towers saw a cloud of dust in the distance as Theoderic approached. Attila was forced back in the direction of Troyes to consider his options. According to one account, he was attacked in the rear while retreating and took heavy losses. But the allies were now united, with Theoderic and Aetius cooperating splendidly and all the allied contingents in place. All were agreed that they had to collaborate or perish separately. Historians cannot agree on a date for the battle that followed, but it was probably 20 June. Attila found terrain where he thought his bowmen would be effective. The exact location of the battlefield is unknown – it is referred to in the sources as the Cataulanian Plains or Fields – but is thought to have been in the environs of Châlons. On the evening of the 19th, then, Attila consulted his shamans. The soothsayers examined the entrails of slaughtered cattle. They predicted disaster but said that an enemy chief would certainly be slain. Thinking this must mean Aetius, and worried that the garrisons on the line of retreat would maul him badly if he retreated now, Attila drew up his forces and exhorted them for the ordeal to come on the morrow.

THE BATTLE OF CHÂLONS OR THE CATAULANIAN FIELDS

On 20 June both sides braced themselves for the fray. Despite loose talk of Hunnic 'hordes' it is likely that the allies had a clear superiority in numbers. There is no scholarly agreement on how many fought, but there must have been at least 30,000 on Attila's side, even granted his (unknown) losses the previous day. The allies must have had more than this, perhaps 40,000 in all, for Attila was clearly worried that in a dawn-to-dusk encounter, with the emphasis on attrition, enemy numbers would tell. We may infer that Attila was nervous about the outcome of the contest, for he delayed giving battle until 3 p.m., doubtless calculating that in the case of his defeat, the enemy would not have enough daylight left to complete a rout. His best-case scenario was a quick victory with the help of the archers, then nightfall, and then the chance to finish off next day. He was relying on his archers to break the allied centre, and indeed the

Romans seemed to have given hostages to fortune by stationing the unreliable Alans there.

Attila drew up his army with his élite Hun units in the centre, the Ostrogoths on the left facing their cousins the Visigoths and the Gepids facing the Romans and others on Attila's right. On the allied side Aetius commanded the left wing and Theoderic the right, with Sangibanus, whose loyalty was suspect, in the middle where the two principal commanders could keep an eye on him. The ancient sources are vague on the details of the battle, but the early stages involved a struggle for a ridge dominating the battlefield, ascended by a steep slope. All authorities agree that the initial fighting was ferocious, but the allies soon won the struggle for the ridge, from which they launched a counter-attack where the Visigoths scored a surprisingly easy victory against Attila's élite regiments. We hear of Attila desperately pleading with his men to remember their valour of yore, and then of a stream in the middle of the plain running red with blood. It was in the course of this heavy fighting that Theoderic was slain, but again the details are vague. One version has him falling off his horse and being dragged to his death, another, more dubious, speaks of single combat between him and an Ostrogoth champion.

By nightfall the Huns had been driven back into a defensive circle of wagons. Short of food and water, they were in serious danger of annihilation. That night Halley's Comet was visible, and the superstitious Attila put this omen together with the previous warning from his soothsayers and imagined disaster. He told his men he was prepared to perish on the spot and made a funeral pyre of wooden saddles, ready to sacrifice himself in the morning. It was a gloomy night, and both Thorismund, Theoderic's son, and Aetius at different times nearly blundered into the Huns' camp in the darkness. Morning came and Attila prepared for the worst. But apart from the routine showers of arrows, there was no move from the allies. Attila concluded the enemy were preparing to starve him out, which they could easily do. But as the day

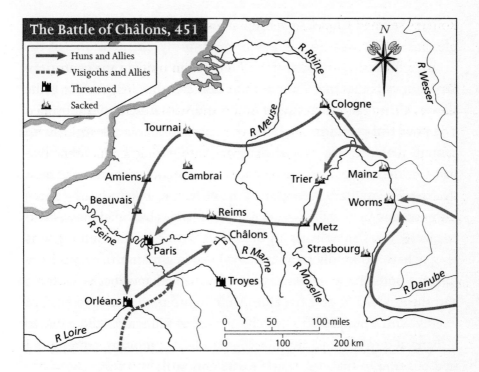

The Battle of Châlons, 451

Huns and Allies
Visigoths and Allies
Threatened
Sacked

N

R Rhine
R Wesser
R Meuse
Cologne
Tournai
Cambrai
Amiens
Trier
Mainz
Beauvais
Worms
R Seine
Reims
Metz
Châlons
Paris
R Marne
Strasbourg
Troyes
R Moselle
Orléans
R Danube
R Loire

0   50   100 miles
0   100   200 km

wore on there was still no stirring from the allies; eventually Attila's scouts brought word that the Huns were not encircled, as feared, but could easily make their escape. Cautiously, Attila began the retreat, at first thinking Aetius meant to gull him with a feigned retreat. But there was no allied follow-up. Aetius had decided that the total destruction of the Huns did not suit his purpose, for it would make the Visigoths too powerful. Acting the Dutch uncle, he advised Theoderic to bury his father with elaborate funeral rites and then return to the Visigoth capital of Toulouse, to prevent his brother seizing the throne. Theoderic would have seen through Aetius's game, but the callow Thorismund either did not do so or lacked the stomach for another day of slaughter. The 20th had certainly been bloody, with the dead numbering at least 15,000 in gut-wrenching close combat. The obliging Aetius left Attila a clear route out of Gaul. By autumn Attila was back where he started, on the Hungarian plain.

107

The battle of Châlons or the Cataulanian Fields was a great feat of arms but it was not, as has been absurdly claimed, one of the decisive battles of world history. One of the oldest axioms in military history is that one should try to beat the opponent before battle is even joined, and this is what Aetius did with Attila in Gaul in 451. The powerful coalition of all the top tribes in France meant that the strategic balance was overwhelmingly with Aetius. Even if the allies had lost the battle, it is unlikely that the outcome would have been decisive. The military resources of the Visigoths, to say nothing of the Burgundians and the Bagaudae were formidable; had not Aetius spent 20 years in vain trying to defeat them? And would Geiseric simply have stood idly by if Attila had become master of France? The obvious outcome would surely have been an alliance between Rome and the Vandals. In 451 the allies were fighting on home ground an enemy with a supply line stretching back to Hungary, vulnerable to a rising at any point along this slender lifeline; moreover, they were in a position to make up battle losses and Attila was not.

The essential truth that Attila always worried about manpower shortages has been obscured by nearly 2000 years of black propaganda about 'the wrath of God' and the idea that hundreds of thousands of Huns were simply flooding into Europe in a kind of unstoppable tsunami. History teaches us to beware Western myths about their armies being swamped by 'oriental hordes', whether the Huns, the Mongols or even the Chinese People's Army in the Korean War of 1950–53. Attila has been overrated, both as a threat and as a general. The truth is that, as a captain, he never beat very much in the military sense, just some second-division Byzantine troops for the most part and some feeble garrisons and sparsely populated towns. He won a pyrrhic victory against the eastern empire in 447 and was soundly beaten at Châlons in 451. It is not much on which to build such a gigantic reputation.

THE ATTACK ON ITALY

Over the winter of 451–2 Attila pondered his options. In a single year he had dented his credibility and shown clearly that he was very far from invincible. Not much had been brought back from western Europe in the way of loot, and unless the leader was soon to make considerable amends for his failure, his position would become precarious. His decision to campaign against the western empire in Italy in 452 hardly came as a surprise. He badly needed a return match against Aetius that he could win and calculated that in Italy Aetius would not be able to count on the power of the Visigoths and the Bagaudae. Furthermore, a foray into Italy might make it possible to revive the moribund alliance with the Vandals. If Geiseric learnt that he was advancing down the peninsula, he would surely send his navy against Rome. Once they had defeated Rome together, the two great leaders could divide the spoils and come to an agreement that would enable them to gain the truly great prize: Constantinople. Most of all, Attila could not simply skulk in Hungary and lick his wounds, for the entire system of financial predation he had built up depended on ceaseless conquests and inflows of gold.

There are those who say that the decision to strike at Italy instead of Marcian and the east was a bad call, but from Attila's point of view the feeble armies of Italy, not in the same class as the Visigoths, looked a much more tempting target than the triple walls of Constantinople. There is a further controversy about Roman military dispositions, with some historians alleging that Aetius was unprepared for this counterstroke. It is true that he did not defend the Alpine passes, and has been criticized for this, but the criticism is misplaced. There were too many wide passes in the Alps and blocking them up required numbers that Aetius simply did not have. Moreover, Italy was invaded six times in the fifth century, and the invaders were never halted in the Alps. The likelihood is that Aetius was concerned about the ravages of famine and malaria in Italy, could not risk troops in the far north of Italy, but thought that disease and pestilence might do his work for him.

Yet at first Attila made a very good showing. He swept into Italy and made for Aquileia, where his initial attack was beaten off with heavy losses. His men begged him to call off the siege, but he sent for better siege engineers and equipment. Legend says he was encouraged to continue the investment of Aquileia when he saw white storks leaving their nests on the city rooftops and flying away to the north. Aquileia was notoriously a tough nut to crack – the German tribes had tried to take it in the days of Marcus Aurelius and failed – but this time it did fall, for the first time in its history. Attila razed it and gave his troops full licence to rape and plunder. Tradition says that it was refugees from Attila's sack of Aquileia who found refuge in the lagoon island and founded Venice. Attila's tactics in this campaign soon became clear: the use of terror to cow the population. This time there was no mercy even for the cities that opened their gates to him and did not resist: they were still razed and gutted. A long line of cities was given over to rapine: Concordia, Altinum, Patavium, Vincenza, Verona, Brixia and Bergamo; only Ticinum and Milan escaped the experience of sack. Milan provided another Attila legend. In the palace there he was said to have seen a painting of the two Roman emperors, of west and east, sitting on a golden throne with slain Scythians around them. He then had a similar painting made of himself on a golden throne, with the two emperors pouring gold at his feet.

The obvious course now was for Attila to turn south and march on Rome. Instead he temporized, marching back east to Lake Garda, keeping his options open. Weighing on his mind were the losses his troops were already taking from the malaria epidemic. There had only been one major engagement so far, at Aquileia, yet Hun numbers were shrinking. All his instincts dictated a retreat to Hungary, but before that could happen Attila needed to save face. Perhaps divining his uncertainty, Rome sent him a high-powered delegation, containing the ex-prefect Trygetius, a bird-brained senator named Gennadius Avienus, who had been consul in 450 and, most impressively, Pope Leo, one of the great early churchmen. The envoys met Attila on the banks of the River Mincio at Peschiera on Lake Garda. The king of the Huns, always flattered by attention from grandees, was particularly impressed by the attention paid to him by Pope Leo, whom he termed 'the highest priest'. After talks with the delegates, Attila suddenly agreed to leave Italy. This apparent about-face has sometimes puzzled historians. Christian propagandists like to tell a story that Leo terrified Attila when a vision of saints Peter and Paul appeared by his side. The truth is more prosaic. Three main factors weighed with Attila. Leo brought him a hefty bribe in gold, more than enough to keep the logades quiet. Attila could therefore return to Hungary and claim that he had achieved all his aims. Moreover, Attila's own shamans told him that any alien who entered Rome was under a curse; Alaraic had sacked the city in 410 and died mysteriously soon afterwards. Attila always took the omens and signs interpreted by his witch-doctors seriously. Finally, and probably the most potent factor of all in the decision to retreat, Attila was gravely concerned about the impact of malaria on his army; the stories from the envoys of the appalling situation further south probably lost nothing in the telling. As a final decider, he received word that Marcian had taken advantage of his absence to send an army across the Danube, which routed the reserves Attila had left on home guard duty.

POLITICAL MARRIAGE AND DEATH

Back in Hungary in the winter of 452–3 Attila planned yet another war of conquest. Logically, this time he would have to direct his

111

efforts against Marcian, if only to make the Huns forget the two failures in the west. Attila must have considered his chances of a decisive encounter against the Byzantines very slender, now that he had lost any chance of support from Geiseric's navy, but his options were narrowing. The failure of the two back-to-back enterprises in the west had seriously weakened his prestige at home, encouraging a previously unthinkable opposition to arise. It was in this context that the polygamous Attila made another political marriage, adding a German princess named Ildico to his bevy of wives. The Huns placed a high value on the possession of high-born women, viewing them as an essential annexe to their prestige and frequently demanding princesses and 'ladies' as part of the tribute from conquered foes. An elaborate wedding feast was held, early in 453. Suddenly, Attila was dead. This is how the historian Jordanes reports this sensational event:

> Shortly before he died, as the historian Priscus relates, he took in marriage a very beautiful girl named Ildico, after countless other wives, as was the custom of his race. He had given himself up to excessive joy at his wedding, and as he lay on his back, heavy with wine and sleep, a rush of superfluous blood, which would ordinarily have flowed from his nose, streamed in deadly course down his throat and killed him, since it was hindered in the usual passages. On the following day, when a great part of the morning was spent, the royal attendants suspected some ill and, after a great uproar, broke in the doors. There they found the death of Attila accomplished by an effusion of blood, without any wound, and the girl with downcast face weeping behind her veil.

This is the conventional version of Attila's death, usually accepted by historians. The symptoms described suggest either a burst peptic ulcer, possibly caused by the extreme stress Attila was now living under, or portal hypertension – varicose veins in the throat caused by excessive alcoholism. If one of these veins burst, the blood would go straight into his lungs if he was lying on his back; if he had been upright or sober, the attack would not have been fatal.

The Huns mourned their great leader. This is how Priscus describes the funeral rites:

> In the middle of a plain his body was laid out in a silken tent, and a remarkable spectacle was solemnly performed. For in the place where he had been laid out the best horsemen of the whole Hunnic race rode around in a circle, as if at the circus games, and recited his deeds in a funeral chant ... When they had bewailed him with such lamentations, over his tomb they celebrated with great revelry what they call a *strava* and abandoned themselves to a mixture of joy and funereal grief ... They committed his body to the earth in the secrecy of the night ... They added the arms of enemies won in combat, trappings gleaming with various precious stones and ornaments of various types, the marks of royal glory. Moreover, in order that such great riches be kept safe from human curiosity, those to whom the task was delegated they rewarded abominably by killing them.

Jordanes adds details linking the motif of precious metals with the Huns. 'They bound his coffins, the first with gold, the second with silver and the third with the strength of iron ... iron because he subdued the nations, gold and silver because he received the honours of both empires.' Some have detected an element of 'protesting too much' in the funeral hymn, which listed Attila's achievements and then went on: 'When he had accomplished all this by the favour of fortune, he fell not by wound of the foe, nor by the treachery of his friends, but in the midst of his people at peace, happy in his joy and with no sense of pain. Who can even call this a death, when no one believes it calls for vengeance.'

A NATURAL DEATH?

Yet the official version of Attila's death is problematical on a number of different levels. Veils, weeping brides, no screams, locked doors – it all sounds far too much like a Sherlock Holmes mystery; the pat description is unconvincing. The received version can be criticized on a number of different levels. Priscus, the historian on whom we must rely for Attila, was not even in Europe when the great Hun leader died, being absent in Egypt on a diplomatic mission. His account was second-hand or even third-hand. It bears no relation to Priscus's own first-hand account of Attila at the banquet in 449. There, other Huns are shown drinking and carousing, but Attila is detached, abstemious, sipping where others quaff. In short, the character of Attila presented by Priscus as an observer is very far from the personality revealed in his later description of his death. Naturally, it was in the interest of Attila's sons and the logades to collude in the idea of a sudden, natural death; the last thing they needed if they were to effect a peaceful succession was a lengthy inquiry into an assassination plot.

Some of the difficulties in the official story are obvious, and many relate to the Christian demonization of Attila as an egregious 'scourge' of God. Christians hated Attila, as they never hated Alaric and Geiseric who, though heretics, were still within the cultural fold. Attila, on the other hand, was a real pagan, a believer in magic and witchcraft, a true child of darkness. What better revenge for Christian promoters of the black legend of Attila than to claim that God punished the blasphemous Hun by condemning him to die, not in glorious battle at the head of his troops, not even on the Gotterdammerung funeral pyre he had prepared for himself at the Cataulanian Fields, but in a drunken and lustful debauch in his private quarters? At least three different well-known stories seem to have provided elements for the official version of Attila's death. There is the biblical story of Judith and Holofernes, where a woman is the agent of destruction. There is the story of the previous Hunnish king Octar, whose death is described in similar terms. And there is the famous denunciation of drunkenness in one of St John Chrysostom's homilies, which could almost have been lifted to supply the details of Attila's fatal haemorrhage. The official story of Attila's death, then, was not a naturalistic account but a moralistic one, designed to point a moral and teach a lesson, which was why it was taken over so avidly by Chaucer in *The Pardoner's Tale*.

If the version provided by Jordanes and Priscus is suspect, what are the likely circumstances of Attila's death? Here the finger of suspicion quite clearly points back to Marcian, an implacable enemy whom Attila had woefully underrated. The Byzantines were quite prepared to use assassination as a political tool, as the Edeco–Chrysaphius plot of 449 shows clearly. Marcian's motive was clear, but did he have means and opportunity, especially given that Attila, on guard against a murder attempt, constantly rotated his bodyguards so that they could not be suborned into a plot; for how can you promise to assassinate somebody if you don't even know if you will be on duty? Some have speculated that the Byzantines may have made over Orestes and Constantius, who had Roman backgrounds and maybe residual Roman loyalties. Yet the most intriguing suggestion is that the mastermind behind any plot could have been the very same Edeco who blew Chrysaphius's 449 hit by divulging it to Attila.

The trails of circumstantial evidence are ingenious, and two possible motives for Edeco have been suggested. One is that, as a German, he was vulnerable to the lure of money and gold in a way a Hun, with a non-venal culture of money as power, not an end in itself, would not have been. The other is that Edeco was a loyal follower of Bleda, that he planned a long-spanning revenge, and that the death of Attila in 453 had something to do ultimately with the murder of Bleda in 445. Edeco could have been a genuine double agent or he could have been 'turned' by the Byzantines at some stage. The most brilliantly inventive suggestion is that Edeco revealed the first murder plot of Chrysaphius to Attila because he did not think it would work. Having thus proved himself Attila's unimpeachable and trusted confidant, he could then proceed to a far more subtle plot of his own. This reconstruction raises the art of the double bluff to new heights. As to the means of murder, it is most likely that slow poisoning proved fatal for Attila, possibly even a sip-by-sip ingestion at the wedding banquet.

THE COLLAPSE OF THE HUN EMPIRE

There are many reasons, then, for thinking that Attila may not have died a natural death. Certainly, if he did, it was miraculously expedient for Marcian and the Byzantines. But in a larger sense it is probably fruitless to speculate on the exact circumstances of the demise of the great Hun leader. The last hours of many great men remain mysterious: we still do not know the full circumstances behind the suspicious deaths of Alexander the Great, Julius Caesar, Napoleon, Lincoln, Stalin and even John F. Kennedy. What is certain is that shortly after Attila's death, the Hun empire came apart at the seams. Attila had not sensed that his end was imminent and so had not made provision for the succession. Almost inevitably his sons fought among themselves for the crown.

While this civil war raged, the subject peoples took the opportunity to break away. First to go were the Ostrogoths and then the Gepids under Ardaric, formerly close friend and confidant of Attila. In 455 the Gepids and their allies scored a great victory over the Huns at the River Nedao in Pannonia, where the slaughter was said to have been terrific and Attila's son Ellac was killed. The other sons and survivors quit Hungary and fled back to their original homeland by the Black Sea. Having regrouped, they remained confident that they could bring the Ostrogoths to heel, as that tribe had not supported the Gepids at Nedao and would have to fight alone. But the Ostrogoths in turn won a shattering victory over the Huns; this time Attila's son Valamer was slain. His favourite, Ernac, was now the only son left, and he was reduced to asking Marcian's permission to shelter in an enclave at the confluence of the Danube and the Theiss. In 469 the Hunnic leader Dengizich (according to some reports, another of Attila's sons) made a last attempt to regain the former glory by campaigning against the Byzantines but he too was heavily defeated and killed. Dengizich's severed head was publicly displayed in Constantinople as a warning to all who would defy the might of Byzantium. The luckless Ernac was said to have ended his days as an obscure mercenary in the service of Constantinople. In a remarkably short time the Huns were no more than a distant memory, soon to be replaced by the Avars as the great power on the steppes. Reflecting on the 'how are the mighty fallen' theme, one historian summed up like this: 'The steppe was now crowded with military nations, among whom the pitiful remnants of the Huns played nothing more than the role of minor robbers and cattle-raiders.'

By a curious historical congruence, the disappearance of the Huns almost exactly coincided with the end of the western Roman empire. The most interesting internal political feature of the Roman domain in the early 450s was the close alliance between three men: Aetius, Marcian and Pope Leo. While Aetius and Marcian collaborated to ward off the threat of Attila, Marcian and Leo imposed religious orthodoxy at the Council of Chalcedon against the challenge of the Monophysite heresy, which the emperor Theodosius had supported. Some have seen the near simultaneity of the battle of Châlons and the Council of Chalcedon as a sign that the Church itself regarded heresy and the 'scourge of God' as the two horns of a single devil.

But Aetius's entente with Marcian carried its own dangers. Just as he had dumped Attila once he no longer needed him and could get a better deal out of a marriage between his son and Valentinian's daughter, so he may have hinted unguardedly to the western emperor that he no longer needed him now that he had Marcian's patronage. Meanwhile, a shadowy figure named Petronius Maximus, evidently an elderly senator, gained Valentinian's ear and poisoned him against Aetius. In September 454, during an audience with Valentinian, Aetius began arguing strenuously against one of the emperor's propositions. Suddenly Valentinian drew a sword and cut him down on the spot; almost certainly the murder had been pre-arranged by Petronius and Valentinian. But things did not turn out as Petronius expected. Valentinian did not make him his chancellor and instead began to distance himself, preferring the company of a eunuch named Heraclius. In the spring of 455 Petronius then assassinated them.

This was the point when the patient Geiseric finally made his move, after closely observing the amazing events in Europe over the past five years. Whether on his own initiative or at the invitation of Valentinian's widow Eudoxia, he ordered his fleet to sail for Rome. The useless Petronius, very good at murder but not much else, fled in terror. As the Vandals advanced inland from the mouth of the Tiber, Pope Leo went out to meet them, hoping to achieve what he had with Attila. But Geiseric was a tougher customer altogether. Leo got only half of what he wanted. The Vandal king agreed not to massacre the inhabitants of Rome or to raze it, but otherwise he ordered a most thorough sack of the eternal city, lasting a full two weeks. This occupation was quite unlike Alaric's relatively innocuous one in 410. At least two-thirds of all valuables and treasures were removed, including entire jewelled rooftops, and all the surviving members of the royal family were carried off to North Africa. With the western empire in limbo, Aetius's old ally Avitus was recognized as the new emperor by the Visigoths, who increasingly emerged as the arbiters of Rome's imperial destiny. Avitus himself struggled valiantly with an impossible situation but lasted barely a year. Seven more phantom emperors came and went before the Goth Odoacer finally rang down the curtain on the western Roman empire by deposing Romulus Augustulus in 476. Rome and the Huns disappeared virtually at the same time, just as Attila and Aetius, bound together in so many ways in life, perished within 18 months of each other. As a final irony, Odoacer was the son of the very Edeco who may well have compassed Attila's death.

THE HUNS AND THE FALL OF THE ROMAN EMPIRE

The exact role of the Huns in causing the fall of the Roman empire in the west will always be controversial. On the one hand, it could be said that the steady stream of Hun mercenaries provided to the Romans by the kings before Attila kept the Visigoths and Burgundians at bay and prevented Germanic tribes dismembering the empire; on this view it could even be said that the Huns delayed the fall of the Roman empire. On the other hand, the massive shunt by the Huns in the 370s probably pushed the German tribes into the empire deeper and sooner; Alaric, the early scourge of Rome, was so troublesome simply because the Hun presence made it impossible for him to return to Germany. A mixed verdict might be that if the Huns retarded the decline of Rome in the years 430–55, they had already accelerated it before that date. And how does Attila himself relate to all this? Some points are clear. The threat from Attila meant the two halves of the empire could not concentrate on the Vandals and the reconquest of North Africa. The western empire could not deal decisively with the Suevi and Visigoths in Spain, and so lost a tax-rich province. Also, the Romans were forced to pull out of

119

Britain definitively, since Aetius could not answer the last desperate pleas for help from the Romano-British kingdom, beset by the Saxons.

But none of this made Attila a towering figure. The case for his greatness can hardly rest on his (mediocre) military performance or his (nonexistent) diplomatic talents. It can only be that he alone realized the potential of Hun society, that he realized, as no one had before, that a Hun confederacy could dominate Europe. He based his power on vassals like Onegesius, Berichus and Edeco, bound to him personally with an allegiance that transcended tribal obligations. In other words, he tried to replace tribalism with feudalism. That this confederacy collapsed immediately after his death does not of course mean that the experiment was doomed to failure and, by the same token, we cannot legitimately say that only the genius of Attila kept it going. Nomad empires had the potential to be long-lived, as the Mongols later proved. Napoleon used to lament that his brothers were feckless, and that Genghis Khan was lucky in having four talented sons. Attila had the sons that Napoleon lacked but, sadly for him, they were not major figures like Genghis's progeny. Attila had the right number of sons, but they proved to be as useless as Napoleon's brothers.

CHAPTER 3

Richard the Lionheart

England's greatest warrior-king

WHEN RICHARD I, king of England, nicknamed Coeur de Lion, disembarked in the Holy Land on 8 June 1191, he was already one of the most famous warriors in Western Europe. Aged 33, he was a veteran of 18 years of almost constant warfare, which began in 1173 when he was not quite 16. Before he left England in December 1189 to begin his famous exploits in the Third Crusade, critics might have been able to say that, despite his great talent at lightning warfare and especially siegecraft, he had not really been tested, since he had never fought a pitched battle and all his experience had been on limited campaigns in France. Most of these interminable wars between England and France were fought in the no man's land of the Vexin in northeastern Normandy. In the twelfth century the Angevin dynasty that ruled England also controlled an informal empire embracing Brittany, Normandy and most of western France as far south as the Pyrenees. France, meanwhile, was a small kingdom centred on Paris and just beginning its rise to greatness under a determined king, Philip Augustus. The campaigns in the Vexin had always ended inconclusively.

A WARRIOR FOR ALL SEASONS

Yet in the past 18 months Richard the Lionheart had proved himself a warrior for all seasons. Unlike his father and brothers, he was sincerely devoted to the cause of Christianity in the Holy Land and had 'taken the cross' even before he became king, swearing a mighty oath to be God's champion against the heathen Muslims. Marching rapidly down through France and Italy, he intervened in civil war in Sicily to back his sister Joan, who was on the losing side – she had been married to William II but when he died his dynastic claims had been set aside by a pretender. After rapidly defeating the forces of the new Sicilian ruler, Tancred of Lecce, Richard sailed to Cyprus, became embroiled in another war there and deposed the self-styled 'emperor' Isaac Comnenus. Any idea that Richard was simply a technician versed in small-scale French campaigns was already unsustainable.

Of all great warriors in history, Richard certainly had the most illustrious parentage, since his father and mother were both titanic figures. Henry II of England was by the 1170s probably the most powerful and successful monarch in Europe, while his wife (Richard's mother) Eleanor of Aquitaine was already a contemporary legend: indefatigable intriguer, accomplished politician, patron of troubadours and even veteran of the Second Crusade. Eleanor had borne five sons to Henry, but by 1191 only Richard and his younger brother John (then aged 24) survived. Two of those who had died young, Henry (the so-called Young King), the apple of his father's eye, and Geoffrey, Count of Brittany, were so notorious for excess that Henry II's family was already known as the Devil's Brood. In the confused and shifting kaleidoscope of Angevin alliances, Richard had fought both with his brothers against his father and against his brothers. Refusing to compromise over the maternal fief of Aquitaine, Richard finally defeated his father, who died cursing him.

He then released his mother who had been imprisoned by Henry since she fomented the civil war of 1173 that set the Devil's Brood against their father. Crowned king of England and overlord of the rest of the Angevin empire in 1189, Richard set out on crusade; he was uninterested in England and would spend only six months of a ten-year reign there. Yet he risked grave unpopularity through soaking the country for money for the crusades, instituting the so-called Saladin tithe. The decree establishing this read as follows:

> Each person will give this year in alms for the aid of the land of Jerusalem a tenth of his income and movable goods, except for his arms, horses and clothes in the case of a knight and any sort of furniture in the case of a cleric; and except for precious stones belonging to both clergy and laity ... but clerics and knights who have taken the cross will giving nothing of their tenth, except what they give for their personal property and for their demesne lands, and whatever their men have owed must be collected for the crusaders' enterprise.

When Richard arrived at Acre in June 1191, the Christian kingdoms of Palestine, established after the First Crusade and therefore nearly one hundred years old, had been fighting the Saracens under Saladin, vizier of Cairo and Damascus, for four years. Aged 50 when he began his Holy War against the infidels, Saladin had spent his life subduing rival Arab leaders in Egypt, Syria and Iraq, thus delaying his longed-for goal of jihad. Although not the supreme Saracen power in the Middle East, since the caliph of Baghdad, his nominal superior, was jealous of him and refused to aid him militarily, Saladin had the advantage over the Christians of a unified command, whereas the heirs to the crusader kingdoms were riven by factionalism, with the lords of Jerusalem, Antioch, Tripoli and Kerak all claiming supremacy. The other disadvantage the Christians suffered was shortage of manpower. Only 250,000 of them held down Palestine and the coast of Syria, being scattered in the inland cities of Jerusalem, Tiberias, Antioch and Edessa, the coastal cities of Latakia, Tortisa, Tripoli, Beirut, Tyre, Acre, Caesarea, Haifa, Jaffa and Ascalon and in the famous crusader castles, which numbered about 50. With an army of 25,000 Saladin engaged a Christian host of 20,000 at Hattin in 1187 and annihilated it. Saladin then swept over crusader Palestine, taking Jerusalem and most of the Christian cities of the coast. Hearing that the holy city was lost to the Saracens, Pope Gregory VIII issued an encyclical calling on the faithful in Europe to go on crusade to help their co-religionists.

UNDER THE BANNER OF THE CROSS

No subject is more controversial than the crusades, and forests have been felled to make the books that would explain the phenomenon. There have always been those who view them simply as large-scale plundering expeditions, legitimized by the sanction of Holy Mother Church. The German philosopher Friedrich Nietzsche's withering assessment is well known: 'Superior piracy, that is all'. The crusades functioned as a kind of out-relief for penniless adventurers, dispossessed nobles and the younger sons of aristocratic families who were the victims of primogeniture, where all the family property went to the eldest son. They also acted as a magnet for mercenaries and get-rich-quick soldiers of all types. Yet there was always more to it than simple greed, for the materialistic motives were overlaid by a fanatical religiosity of a kind that is always impossible to imagine or retrieve in the twenty-first century. It has been well said that we can no more understand the crusades than a medieval person could understand landings on the moon. The best way to envisage the crusades is as a crude mélange of greed and piety, and there is a

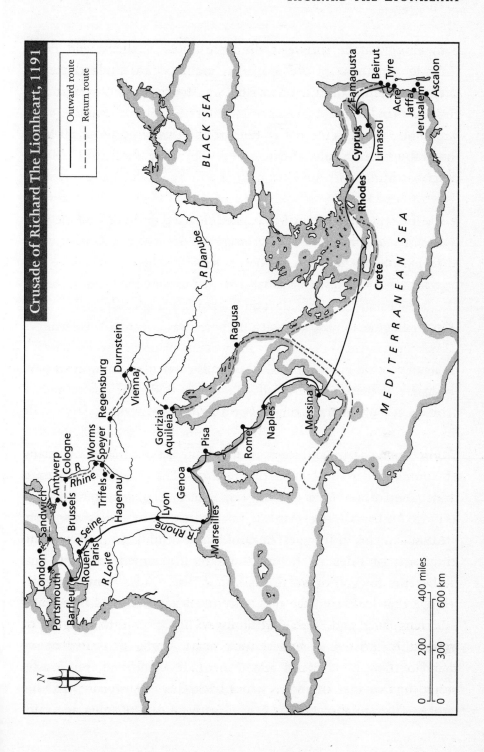

Crusade of Richard The Lionheart, 1191

Outward route
Return route

BLACK SEA

MEDITERRANEAN SEA

Famagusta
Beirut
Tyre
Ascalon
Acre
Jaffa
Jerusalem
Cyprus
Limassol
Rhodes
Crete
Ragusa
Durnstein
Vienna
Regensburg
Gorizia
Aquileia
Pisa
Rome
Naples
Messina
Genoa
Lyon
Marseilles
Worms
Speyer
Trifels
Hagenau
Cologne
Antwerp
Brussels
Sandwich
London
Portsmouth
Barfleur
Rouen
Paris

R Danube
R Rhine
R Seine
R Loire
R Rhone

N

400 miles
600 km
200
300
0

compelling parallel between the motives of those on the Third Crusade and those of the Spanish conquistadores three centuries later. No one on the Christian side was more sinister, cynical and ruthless than Conrad, Marquis of Montferrat, a 'native' crusader who would be a thorn in the side of Richard the Lionheart. Yet this is how the Arab historian Baha al-Din, an eyewitness of the Third Crusade, describes him:

> He had a picture of Jerusalem painted showing the Church of the
> Resurrection, the object of pilgrimage and deepest veneration to
> them; according to them the Messiah's tomb is there, in which he
> was buried after his crucifixion ... Above the tomb the marquis had
> a horse painted, and mounted on it a Muslim knight who was
> trampling the tomb, over which his horse was urinating. This picture
> was sent abroad to the markets and meeting places; priests carried it
> about, clothed in their habits, their heads covered, groaning: 'O the
> shame!' In this way they raised a huge army, God alone knows how
> many, among them the king of Germany with his troops.

Troops, mercenaries and soldiers of fortune set out from all corners of Europe under the banner of the Cross. There were contingents from Genoa, Pisa, Venice, Hungary, France, England, the Angevin empire, Austria and Germany. If Saladin had been one of the very greatest captains of the ages, he could have annihilated the Christian kingdoms of Palestine before these reinforcements arrived. But he became bogged down in a siege of Tyre – a near impregnable fortress that had taxed the ingenuity of Alexander the Great. When this siege failed and Saladin withdrew to lick his wounds and think again, the Christian defenders took heart. Under their two commanders, Guy of Lusignan and Conrad de Montferrat, previously rivals, the Crusader kingdoms struck back. Guy of Lusignan, in a fine stroke of imagination, raised a new army of 10,000 men and laid siege to Acre. Saladin took him in the rear, squeezing Guy's army between the defenders of Acre and his own force, but the Christians fought well and threw the Saracens back. Meanwhile, the first of the Christian reinforcements started to arrive. Alarmed by the thought that German emperor Frederick Barbarossa was marching overland through Turkey with an army of 25,000 to arrive in his rear (the crusades had their share of black comedy), Saladin remained on the defensive throughout 1190 and let the siege of Acre take its course. As more and more reinforcements poured in to supplement Guy of Lusignan's besiegers, Saladin began to despair. His hopes momentarily lifted when he heard that Frederick Barbarossa had drowned in a river in Turkey and his army dispersed, but he sank into gloom again when he heard of the coming of the western kings. Philip Augustus of France, an old enemy of Richard's, arrived in April 1191 and then, two months later, came the Lionheart himself.

THE LIONHEART IN PALESTINE

The psychological effect of the coming of a king with such a great reputation as a warrior was the last straw for the Islamic defenders of Acre. Although they had resisted valiantly for 21 months, they collapsed almost overnight once the Lionheart set foot in Palestine; four days after his arrival, they surrendered. This virtual walkover victory enhanced Richard's prestige, but he had no illusions about the difficulty of the main task before him: the recapture of Jerusalem. Pre-eminent among his problems was the besetting sin of Crusader Palestine: excessive factionalism. Guy of Lusignan and Conrad de Montferrat remained deadly rivals for supremacy even though they had collaborated. Complicating issues immensely were the two independent military orders of knights, the Templars and the Hospitallers, who fought with and for the Christian rulers but had their own agenda. Philip of France was intriguing with Conrad against Guy of Lusignan and was insanely jealous of Richard's easy victory at Acre; he always hated the Lionheart but he construed the Saracen surrender almost as a personal insult to himself. When Richard persuaded Conrad to accept Guy as overall ruler of Christian Palestine, with the proviso that his (Conrad's) heirs would succeed, thus blunting Philip's intrigues, the French king became incandescent with rage and abandoned the crusade forthwith, leaving part of his army but making all haste to return to Europe. Richard had not been tactful with Philip and the French, but he compounded his lack of diplomacy by alienating the German commander. When Frederick Barbarossa was drowned, command of the German contingent passed to his son, Duke Frederick of Swabia, but he in turn died at Acre and the command devolved on Duke Leopold of Austria. Leopold claimed that Richard insulted him and he too stormed off and took ship for Europe. Soon both Philip and Leopold were spreading propaganda around Europe to the effect that Richard was a monster, an unChristian assassin and poisoner. Since Richard was in Palestine he could not answer the charges and the absent, as the French proverb tells us, are always wrong.

The upshot was that Richard began the campaign to retrieve Jerusalem in command of a sullen and demoralized army, many of whom owed their allegiance to kings and lords who had departed. Old enmities were not forgotten either, with the Pisans and Genoans at each other's throats, and the Hospitallers and Templars jealously guarding their privileges and prerogatives. Nor could Richard, who sincerely believed in the values of chivalry, look for much from his own men. Remarkably similar in motivation to Cortés's conquistadores, they had come to the Holy Land to become rich and grow fat on rapine and booty; the Cross merely served as a fig leaf for their low ambitions. While he waited for the terms of the surrender at Acre to be implemented, Richard reviewed the heterogeneous forces

Gladiators at Capua wore helmets similar to this one found at Pompeii.

A gladiator in action, displaying the famed athleticism. Gladiators could be of many different kinds: the Samnites, Thracians and Gauls; the 'retiarius' (net man); the 'secutor' (chaser); the 'mymyrillo' (fish man); and 'velites' (gladiators who fought only on foot with spears.

SPARTACUS

A Samnite gladiator closes in for the kill in this mosaic. Spartacus, though a Thracian by birth, was not a Thracian gladiator but a Samnite.

The perils of the arena. This wall painting from Pompeii shows the riot in the ampitheatre there in 59.

A 'retiarius' (net man) bites the dust. This detail of a mosaic depicts a gladiator fight from Torre Nuova.

Kirk Douglas as Spartacus in the 1960 film. With characteristic comic hyperbole, Peter Ustinov, who played Batiatus, owner of the gladiator school at Capua, claimed that it took longer to make the film than for the Romans to defeat Spartacus.

Christian princes tremble
before the Hun. Attila was
the first savage chieftain to
conduct a mafia-style
shakedown operation
against the Roman empire:
demanding money with
menaces.

Jack Palance as Attila in the
1954 movie *Sign of the Pagan*.
One in a notable series of
portrayals of psychopathic
villains from the ancient
world by Palance, including
the ace gladiator, Torvald,
in *Barabbas* and Ogodei,
son of Genghis Khan, in
The Mongols.

St Peter and St Paul help Pope Leo to turn Attila away from Rome in 452. In reality, cholera, not an apparition in the sky, was the main deterrent.

A coin showing the empress Galla Placidia, the woman who so fascinated the psychologist C.G. Jung.

Attila's fatal wedding night portrayed in this miniature. How he really died remains an historical puzzle.

Saracens versus Crusaders. The two sides were evenly matched in a military sense, though the Crusaders' heavy cavalry gave them a slight edge.

Sea power: the key to crusader success. Richard both bottled up the Saracen fleet and re-supplied his army from Cyprus.

Saladin: the Saracen military genius proved no match for the Lionheart.

Butchering the prisoners at Acre. Richard's 'war crime' was bizarrely replicated 600 years later by Napoleon, in almost the same place and the same circumstances – perhaps the clearest example of history repeating itself.

Miniature showing the siege of Acre. Richard's military genius manifested itself in a quick end to the 18-month siege.

Richard in full triumphalist pose. There is a certain irony in locating the statue of England's despotic warrior king outside the Palace of Westminster, supposedly the cradle of democracy.

at his disposal. From England and Normandy Richard had brought 9000 men (900 knights, 8100 foot); the Frenchmen left behind by Philip Augustus numbered some 7000 (700 knights, 6300 foot); then there were the men of Guy of Lusignan and Conrad – about 2000, including 20 knights, plus about 1000 Hospitallers and Templars. To these he could add the Scandinavians, the Pisans and Genoese, a Hungarian contingent, a handful of Germans (most had gone home with Leopold) and a sprinkling of mercenaries. When he marched on Jerusalem Richard could pit an army of 20,000 against Saladin's force, perhaps 25,000 strong. The technological gap between the two armies was slender, and long gone were the days of the First Crusade when mounted knights could make mincemeat of enormous Saracen armies of ill-trained levies.

But first Richard had to settle with Acre. The city had been surrendered on the following terms: the defenders were granted their lives provided they converted to Christianity and that Saladin released 1500 prisoners and paid 200,000 dinars for the ransom of the Acre garrison. Saladin, who did not have the money for the ransom, tried to stall and prevaricate. At first he offered to release all his prisoners, give hostages and pay half the ransom provided all the garrison was released – a blatant rewriting of the surrender terms. Then he raised an endless series of cavils and caveats, playing games of logic-chopping and legalese. As the talks dragged on into the second half of August, Richard realized that Saladin was simply playing a game of bluff, hoping to frustrate and demoralize the crusaders who were eager to march south. He became more and more angry with Saladin and finally his patience snapped. On 20 August he ordered all the Saracen prisoners who were not wealthy enough to arrange individual ransoms to be taken out on to the plain outside Acre and slaughtered.

Some 3000 Muslims died that day in an orgy of butchery that has aroused controversy ever since. Richard was criticized for not accepting Saladin's revised offer, but he calculated that once he retrieved the 3000 prisoners, the vizier would welsh on the deal and not pay the second instalment of the ransom money; true, Richard would then kill the hostages but he would simply slay dozens when thousands had escaped. His official reason for the massacre was that Saladin had reneged on the surrender terms, leaving Richard with the option of either letting the garrison go free or having 3000 useless mouths to feed. Some Christian chroniclers justified the killing as revenge for the many crusaders killed at Acre and the liquidation of all the Hospitallers and Templars Saladin had taken prisoner after the battle of Hattin. Others said that letting the garrison go free on parole – which meant nothing to Saracens – simply meant that Richard would have to fight them all over again. Another view is that Richard was impressing his own troops with his ruthlessness and at the same time sending out a message to Saracen garrisons in the coastal cities about the fate that awaited them if they opposed him.

The massacre permanently tarnished Richard's reputation and, in the short term, led to savage reprisals from Saladin, who ordered all Christians captured thereafter to be put to death. But Richard told his captains he was pleased that he did not have to detach any of his troops to guard useless prisoners in Acre. As it turned out, Acre was a threat to his manpower in quite another sense. The city was notorious for its brothels, stews and other low dives, and the fleshpots proved an irresistible draw to the crusaders. Noting that the muster roll of his army was well below putative battalion strength, Richard sent his military police into Acre to roust out the malingerers. The French were the worst scrimshankers, some of them boldly denying that Richard had any authority over them and claiming they would answer only to the senior French officer, Hugh of Burgundy. Gradually, Richard winkled them out of their gambling dens and cathouses, shaming the devout with prayer, cajoling those susceptible to flattery and in some cases simply using force, but it was said that he had to bribe the truly hard cases to return to their regiments. Some of those who lolled in Acre, it must be said, were quite keen to march on Jerusalem; the sooner the Holy City was taken, they reasoned, the sooner they could fulfil their vows as pilgrims and return home laden with loot. Richard and his commanders, using a Judaeo-Christian reflex of misogynism as old as Genesis, blamed the female sex for the anarchy in Acre. So the Lionheart decreed that the strict rules of the crusades would be adhered to on the march south: there would be no women other than washerwomen, for only thus could discipline be maintained and the pressure on food supplies eased. To show that there were no exceptions he made his wife Berengaria and his sister Joan remain behind also.

A BRILLIANT COMMANDER AND MILITARY GENIUS

After beating off a few attacks by the Saracens, the crusader army began the march south. Already a master of skirmish, ambush, lightning probes and siegecraft, Richard now proved himself a brilliant commander in terms of supply lines, communications and logistics. From England he had brought 216 ships on an 8000-km (5000-mile) voyage (his fleet travelled via the Atlantic, the Straits of Gibraltar and the length of the Mediterranean while he proceeded overland) and cleared the Levantine shores of enemy shipping. Crusader supremacy at sea was so overwhelming that some have seen Saladin as a kind of early version of Napoleon, frustrated by the enemy's sea power. The conquest of Cyprus and the control of the sea lanes meant that Richard had an uninterrupted supply line from Cyprus to Acre and thus the possibility of adequately provisioning his army for the long southern march. His logistical task was to ensure that the motley crusader force arrived intact before the walls of Jerusalem, adequately fed and watered, and to keep his lines of supply and communication open while the army lumbered slowly towards its objective; lightning warfare was out, for the crusaders had to convey heavy equipment and siege engines, which would be in constant use during the campaign. These siege engines were of two kinds, the ballista and the mangonel, but it was the ballista on which Richard now concentrated, just as the mangonel had been the weapon of choice at Acre. The mangonel was a swing-beam machine that unleashed stones or other projectiles by using a heavy counterweight. But the ballista, a lighter engine, was better designed for mobile warfare; it could be moved around easily, unlike the mangonel, which was too heavy to be transported in one piece and had to be dismantled and then reassembled.

Nothing better illustrates Richard's military genius than his attention to the most minute details of logistics and commissariat. The first thing to ensure was that each of the 20,000 troops going south had ten days' supply of food and water before setting out. Each soldier carried a 15-kg (33-lb) pack of provisions – 1 kg (2 lb) of food and ½ kg (1 lb) of firewood daily; the timber was essential for lighting fires and boiling water. The normal crusader diet was hard, dry biscuit, a soup of beans and a little salted pork or bacon, supplemented by fresh fruit, vegetables and horsemeat. The 15-kg (33-lb) pack was carried in addition to the infantryman's usual complement of helmet, hauberk, sword, shield, eating utensils and extra clothes. A water ration of 15 litres (4 gallons) a day per man was carried in animal skins by porters or in barrels on horse-drawn carts. Normally, an army could expect to replenish water supplies from streams and rivers, but Richard expected there would be few of these in the desert and that the Saracens would have poisoned all the wells. The crusader army's 6000 knights posed special problems, for each horseman additionally needed 7 kg (15 lb) of fodder and 20 litres (5 gallons) of water a day for his horse, and most knights owned three horses. All in all, to accomplish a 20-day march, Richard's army needed 1340 tons of food and 3 million litres (800,000 gallons) of water.

THE MARCH TOWARDS JERUSALEM

The fastest and most plausible route to Jerusalem from Acre was to march south along the coast to Jaffa, then swing east and strike inland. This route had several advantages. Food, supplies and the siege engines could be carried south by ship for the first leg of the journey, and the crusaders needed only to guard one flank, since the Mediterranean protected the right wing; moreover, Saladin could be kept guessing, since it was possible that Richard's objective was not Jerusalem but the heart of the Saracen kingdom: Egypt itself. For three days, during 22–25 August, the Lionheart painstakingly assembled his army outside Acre. Then the trek began, at first in easy stages to the River Belus and the coastal plain, proceeding at no more than 3 kilometres (2 miles) an hour; Richard wanted to give the men time to shake off the effects of a week of debauchery and dissipation and, also, he needed to allow the fleet to keep up and it was battling contrary winds. Only on the third day did he increase the pace of the march to eat up 18 kilometres (11 miles). On the left wing the Saracen army marched in parallel, just out of range. Saladin continually harassed the Christians, usually sending in groups of lightly mounted archers. To ease the pressure, Richard alternated flank duties, 'spelling' his infantry with tours of duty on the dangerous left and the secure right. Between the two flanks of infantry rode the knights, in three divisions and in close formation, stirrup-to-stirrup, the horses nose-to-tail. The vanguard and rearguard were provided by the Hospitallers and Templars, who rotated this duty on a daily basis.

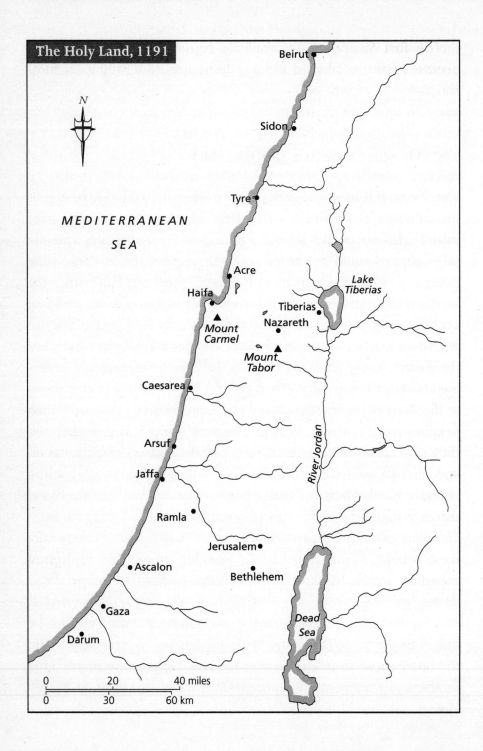

The Holy Land, 1191

N

Beirut

Sidon

Tyre

MEDITERRANEAN
SEA

Acre

Haifa

Lake
Tiberias

Tiberias

*Mount
Carmel*

Nazareth

*Mount
Tabor*

Caesarea

River Jordan

Arsuf

Jaffa

Ramla

Jerusalem

Ascalon

Bethlehem

Gaza

*Dead
Sea*

Darum

| 0 | 20 | 40 miles |
| 0 | 30 | 60 km |

The first Saracen attack, on the rearguard, nearly succeeded because of an error by the Duke of Burgundy who had, unaccountably, left a gap in the formation, whether wittingly or unwittingly; some accused him of simple incompetence but others claimed that, in a sullen rage that he had to take orders from Richard, he had actually sabotaged operations, perhaps on secret orders from Philip Augustus. The Saracens broke through, attacked the wagon train and were on their way to a splendid victory when Richard appeared, having galloped back from the van. After a brisk action Saladin's men were beaten off. The French knight William de Barres, an old enemy of Richard's, so distinguished himself in the fray that Richard publicly forswore the ancient feud and praised the Frenchman lavishly. It was always a point of honour for the Lionheart to be in the thick of action, to uphold the chivalric code by leading by example; he would never ask any soldier to do anything he was not prepared to do himself. For a while the Saracens did not attempt close combat again, but contented themselves with shooting clouds of arrows at the enemy. Fired at such long range, the shafts were rarely fatal but they hung from the crusaders' armour and breastplates, making the Christian troops appear like pincushions or porcupines. The Arab historian Baha al-Din, an eyewitness, wrote: 'I saw some of the Frankish foot soldiers with from one to ten arrows sticking in them, and still advancing at their usual pace without leaving the ranks ... One cannot help admiring the wonderful patience displayed by these people, who bore the most wearying fatigue without having any share in the management of affairs or deriving any personal advantage.'

It took the crusaders 19 days to march the 130 kilometres (81 miles) from Acre to Jaffa. Already, on the first leg to Haifa, they came to realize in what an ordeal they were engaged, even without the Saracens. The temperature reached 40 °C (104 °F) in the daytime – almost unbearable for men in armour and chain mail – the track they trekked along was overgrown with thorn bushes and prickly shrubs, and there were poisonous snakes and tarantulas to watch out for. Men fainted in the heat and, if lucky, were taken on board ship to recuperate; if not, they died where they lay. Everyone commented on the sensory bombardment of a most unwelcome kind. The stench was unspeakable, with the fetor of stale sweat mingling with that of horse dung and human excrement while, particularly at night, the non-stop din was unbearable: not just the cries of the wounded or the snakebitten but the clangour and banging of thousands of pots, pans, shields and helmets, believed to ward off tarantulas and noisome insects. The worst conditions were endured by the professional warriors in the rearguard, for the Templars and Hospitallers had to tread a path churned up by the main army ahead of them, leaving them only loose sand and mud to walk on. Richard tried to ease the suffering of his men by making pre-dawn starts and proceeding only until noon, alternating a day's march with a day or two of rest. So, for example, the 18 kilometres (11 miles) clocked up on 25 August were followed by a rest day, then on 27 August there was a 19-kilometre (12-mile) march from Caiaphas to Athlit, then two days of rest, followed by a 21-kilometre (13-mile) march (the longest ever on the entire campaign) to Merla on the 30th. On this day the Saracens attacked; the dauntless Richard led a cavalry charge to clear the way, but was almost captured. Ambroise, the crusaders' chronicler, suggested that Richard came close to capture through the exhaustion of his own men, as listlessness and apathy were by now assailing the toiling army.

Despite the exertions of the previous day, Richard coaxed his men early next day into completing the 5 kilometres (3 miles) to Caesarea, which would make a convenient base. They found the city in ruins, the battlements demolished by the Saracens before they departed. Richard pitched camp close to the so-called River of the Crocodiles, which flowed into the sea some 5 kilometres (3 miles)

from the gutted town. Here the crusaders received some useful reinforcements from Acre, the final shirkers who had hidden in Acre
when the army departed but who had now been run to earth by
the military police. Richard decided it made more sense simply to
enrol them in the army than punish them, and his edict in Acre
spoke of amnesty: 'The lazy people who were loitering there [Acre]
should by royal command get on board the ships which he had sent
them and come to the army for the love of God and to exalt the religion of the Christian faith and to discharge their vows of pilgrimage
more fully.'

The reanimated army, morale boosted by the newcomers, set
out at 3 a.m. on the morning of 1 September, but ran into the stiffest
Saracen opposition so far in accomplishing the 5 kilometres (3 miles)
to the Dead River. In one heavy skirmish, a notable emir and Muslim
paladin, 'Ayaz the tall', was killed. Already cast down by the fierce
fighting that day, the crusaders allowed themselves to be further
depressed to find the Dead River almost invisible under a dense mat-
ting of reed and rushes – which, in their heat-induced paranoia, they
construed as deliberate Saracen camouflage; it was not, it was a natural feature.

On Tuesday, 3 September, the crusaders found the way ahead
so impenetrable that they diverted inland to find a road running
parallel to the coast. Taking their chance, the Saracens attacked in
force and kept up the impetus during the Christians' 11-kilometre
(7-mile) trek to the Salt River. The Templars took the brunt of
the attack in the running battle and lost a number of horses, as did
a group of French rescuers under the Comte de St Pol, who was
slightly wounded by a crossbow bolt. Richard's men were in glum
mood that night and, to make matters worse, food was beginning to
run short, even with the resupply by sea on 29 August. Sensing the
plummeting morale and hearing of soaring food prices on an informal black market, Richard evinced his talent for man-management

by announcing there would be fresh meat for all. He made good his boast by persuading his knights to kill sufficient horses for the meat ration, paying them sky-high prices for the sacrifice of their steeds. To gain a breathing space, he put out peace feelers to Saladin on 4 September, while the army rested. Next morning, as his men set out on another gruelling march to the River Rochetaille, Richard went out to talk to Saladin's brother Safadin, taking Humphrey de Toron with him as interpreter. Nothing significant transpired – it was obvious that both sides were shadow-boxing. Saladin had told his brother to keep stalling until Saracen reinforcements came in, while Richard made the impossible demand that Saladin return every last Christian possession that he had conquered. Meanwhile, the crusaders had come through the obstacle of the forest of Arsuf – very nervously as there was a strong rumour the Saracens intended to torch it and trap the enemy in an inferno. Nothing happened, but when they emerged from the trees on to the plain of Arsuf, they found Saladin's army drawn up in formation and ready to give battle. The crusaders pitched camp on the north bank of the River Rochetaille, within sight of the enemy.

A FULL RANGE OF MILITARY GIFTS

Richard knew that the coming battle would be a contest between his heavily armoured knights and the mounted archers of the enemy. Delivered at just the right moment, a charge of western knights was irresistible, but everything depended on the timing, for if the crusaders got it wrong, the enemy could part before them and then pick them off one by one. Saladin's tactics depended heavily on mounted archers and their ability to stop the knights before they could get under way. Unless delivered at very close range, their arrows could not pierce the dense armour of Christian horsemen, so they aimed instead at the more vulnerable horses, hoping to disable the enemy cavalry before it could even form up for the charge. To prevent this wrecking tactic, Richard used light horsemen (either native Palestinians or Turkish mercenaries) to keep the Saracen mounted archers at arm's length and, as a second line of defence, a heavy screen of infantry using a shield wall, so that the enemy could not close the range on their horses. Behind this wall the Christian horse would form up, ready for the signal to erupt. The intended movement called for the most meticulous judgement, split-second timing and a flair for anticipating and second-guessing the opponent. It is not often that a warrior possesses strategic brilliance, extraordinary administrative and logistical talents and tactical genius, but Richard the Lionheart had the full range of military gifts.

On the morning of 7 September the Christian army moved out, cautiously, crossed the river and followed the coast road, aiming to bivouac at the town of Arsuf by midday. There were the usual three columns, but Richard had all the baggage on the right, nearest the beach and on the left a dense infantry screen commanded by Count Henry of Champagne. This day's order of march saw the Templars in the van, followed by Bretons, Angevins and Poitevins and then the élite fourth division comprising the English and Normans guarding Richard's dragon standard; behind them came the French, and in the rear, in the position of maximum danger, were the Hospitallers. So tightly packed were the troops that Ambroise recalled that if you heaved an apple into the throng you were bound to hit somebody. Saladin's force kept pace with them, watching for the first sign of weakness while Richard, for his part, rode up and down the left wing looking for any flicker of movement from the Saracens. Standing orders had been issued that no one was to break ranks until Richard gave the signal – which was to be a simultaneous trumpet blast all along the army. At 9 a.m. Saladin made his move, sending a hailstorm of arrows thudding into the crusader rearguard. The

Hospitallers responded valiantly but were heavily outnumbered and soon the Saracens were targeting the precious horses with deadly effect. Already struggling in the ferocious heat, the Grand Master of the Hospitallers sent to Richard to ask permission to charge. He refused.

Delighted with the inroads his men were making on the rear, Saladin threw more and more of his men into the struggle with the Hospitallers, who were soon under intolerable pressure, walking their horses backwards as they faced the enemy while trying to ensure that a gap did not open up between themselves and the Christian centre. A despairing Master left his men to gallop up and accost Richard, imploring him to charge to relieve the pressure. 'Lord king,' he said, 'we are being violently attacked. We will be stained with eternal dishonour if we do not dare fight back.' Again Richard refused, making the infuriating reply: 'Put up with it, Master, no one can be everywhere at once.' The Master returned to the rear to find his men on the verge of cracking. The taunts of the enemy horsemen enraged them and all the while the black cloud of arrows formed parabolas in the sky as it descended on them in a deadly shower. Eye-witnesses described scenes of utmost chaos, with the cries of wounded men counterpointed by the throb of drums and the clashing of cymbals, all played out against a physical background of extreme temperature and shimmering heat haze. The Hospitallers sent several more despairing messages to Richard that morning but each time he exhorted them to hang on just a bit longer, until the Saracen horsemen were tired and could no longer ride away from the nemesis he intended to unleash on them. The spirit in the rearguard became mutinous, with many cursing the king; the consensus was that if they had to die, they would prefer to do it charging the enemy than remaining immobile as sitting targets. Finally the Hospitallers could stand it no longer. Two of their number, the marshal of the order and a knight named Baldwin

Carew, broke cover and charged. The other Hospitallers streamed after them, and the French, in the rear of the centre column, assumed the signal had been given and joined in too. This was a critical moment. If the premature charge were not supported, the Hospitallers would be hopelessly outnumbered and surrounded.

RICHARD, THE UNBEATABLE

Sizing up the situation in an instant, Richard ordered a general cavalry charge. First into the thick of action, as always, he was ably supported by the Angevin and Poitevin horse nearest to him. It was not the textbook action the Lionheart had envisaged but luck, that necessary companion of all successful generals, was with him. Growing over-confident and perceiving the crusaders to be over-timid, many of the mounted archers had dismounted to creep closer for a better aim, and these turtles without their shells were simply bowled over by the charging knights, knocked to the ground to be finished off by the Christian infantry coming up fast behind. Saladin could scarcely believe his eyes. Within minutes his glaringly success-ful attack on the enemy rearguard had been converted into disaster. Psychologically, the Saracens were beaten almost instantly, for sud-denly they had to face the thing they had been constantly warned about: Christian chivalry at full tilt on the dreaded destriers (heavy war-horses). Panic simply increased the tally of slaughter, but there was no doubt that the Saracens did panic, as some of them even climbed trees to try to escape their fate.

The chaos of battle was everywhere in evidence and 'friendly fire' predictably manifested itself. One observer said that the dust kicked up by the charge was so thick that the crusaders 'struck out at ran-dom to left and right because they could not identify their own people by sight, and they cut everyone to pieces as if they were enemies because the dust prevented them from recognising their friends'. Once again luck intervened, and Saladin was able to avoid total rout because of crusader misunderstanding. The Norman and English troops who had not been in the first wave (as they formed the reserve) galloped up to join the fray and clustered around the royal standard. The Angevins and Poitevins in the first wave saw the movement and assumed Richard had ordered all cavalry to group around the flag. They therefore broke off the bloody pursuit of the enemy and rode back to the king.

Seeing this unexpected reprieve, Saladin thought he saw a chance for a come-back and sent in his own reserves, the crack troops of his household regiment, hoping they would rally his fleeing levies. For a while there was ferocious fighting as the élite corps of both sides clashed but Richard and William de Barres both led sustained charges that finally broke the Saracens definitively. As the enemy fled the field, Richard directed three more charges to mop up opposition on the field, but kept a clear head. At the far edge of the battlefield were woods, and he gave strict orders that his cavalry were not to pursue the enemy there, in case Saladin had laid an ambush as a final trick. He knew all about the feigned retreat – a favourite tactic of eastern armies – and that Saladin liked to think through a number of battle outcomes and have contingency plans ready. Richard's cool brain in the moment of euphoria and exultant victory was yet another of his qualities as a warrior. He was left in possession of the field – by some canons the classical sign of victory in a battle – where his men counted some 7000 enemy dead; the crusaders had lost barely one-tenth that number.

Saladin was seriously cast down by the defeat and was reported so depressed that he could not eat. One of his emirs told him candidly that Richard was unbeatable, and the Muslims took to referring to him as Melek Richard (Richard, the true king). Beaten in siege-craft at Acre and now in pitched battle at Arsuf, Saladin, whose reputation five years earlier was as the hammer of the Christians, already looked very mediocre indeed. His one hope was that he could manipulate the enemy by playing on their internal differences and factionalism. From this time on he began seriously to target Conrad of Montferrat.

DIVERGENT OBJECTIVES: SOLDIER OR PILGRIM?

After resting on the day after the battle, the crusaders set out again on 9 September and by noon the next day were traipsing wearily into Jaffa; there had been only token resistance from the Saracens along the 18 kilometres (11 miles) from Arsuf. Once again the forces of the Christian coalition entered a ghost town, almost totally demolished by the enemy. They quickly made camp in the surrounding orchards and olive groves and began to unload supplies from the fleet, which arrived shortly after them, thus completing an impressive amphibious operation. The march from Acre had been a notable achievement. For the first time the heterogeneous elements in this multinational force had functioned as a unit, under a single commander. Richard had not only licked his army into a cohesive whole but also achieved a degree of central control and discipline rare in medieval forces. He had not made the mistakes a lesser commander might have made, such as leaving the baggage train to follow the main column, making it vulnerable to interception, and he had achieved impressive coordination between the army and fleet. Moreover, he still had Saladin in the dark about his ultimate destination, and the vizier guessed that it would be Ascalon rather than Jerusalem, possibly with a view to attacking Egypt. He therefore ordered Ascalon demolished also, to the dismay of its inhabitants and of a significant minority of his war council, who thought the action smacked of weakness and desperation. Hearing of Saladin's actions from his ships patrolling the entire coast, Richard at once thought of a lightning strike to take the Saracens by surprise. But suddenly he found that the miracle of a united command was no more.

The divided counsels among the crusaders, never far from the surface, basically derived from the lack of a common aim, or the divergence of secular and sacred objectives. Those like Richard who wanted to campaign on the basis of military realities wished to destroy Saladin by securing the entire coast. But there were many others who stressed pilgrimage and salvation as the aim of the crusade, and that meant making for Jerusalem. In vain did Richard argue for the coastal route and an assault on Alexandria, and stress the great dangers ahead once the army turned inland and could no longer be supplied by sea; he was outvoted and had to acquiesce. Learning that the Christians were swinging inland, Saladin returned to Jerusalem to put its defences in order and instructed his brother to take up a blocking position at Ramla on the Jaffa–Jerusalem road. Satisfied that Jerusalem could withstand a long siege, he then returned to Ramla on 30 September and, four days later, withdrew his army to a new position at Toron des Chevaliers, 16 kilometres (10 miles) southeast of Ramla and about halfway between Jaffa and Jerusalem.

Richard meanwhile was just as busy. Hearing that Acre still contained large numbers of malingerers, deserters and absentees, he stormed up to the city and solved the problem by bribing the owners of the brothels and bordellos to shift their operations to Jaffa, where the majority of the troops were located, taking the ladies of the night with them. The 'desert dodgers' thus saw no reason to remain in Acre and were persuaded to rejoin their comrades in the south. The price Richard had to pay for this was the collapse of his 'no women in the army' order; as Ambroise put it in a rhyming couplet: 'Back to the host the women came/ And plied the trade of lust and shame.' While in Acre, the Lionheart learnt from his spies that Saladin's secret overtures to Conrad were paying dividends: Conrad had agreed with Saladin's suggestion that he should attack Acre while Richard was in the south, and that in return Saladin would give him Tyre and Sidon.

RICHARD, THE THRILL-SEEKER

Richard returned to Jaffa with Berengaria and Joan, most of his fleet and the reluctant soldiers. Unable to be still for a moment, he insisted on going in person on reconnaissance missions and during one of these, on 29 September, he was nearly captured. It was only the fact that a valiant French knight named William de Preaux called out that he was 'Melek Rik' that diverted the Saracens from the real Lionheart. Richard was always severely criticized by his commanders for these pointless gallery touches – and Saladin thought that reck-lessness was the one thing that prevented Richard from being a truly great captain – but he was deaf to their pleas. He thought that dis-plays of personal courage and derring-do were good for the morale of his men and besides he enjoyed it – part of Richard's psychology was that he was a thrill-seeker. He was also irritated that his grand plan for an amphibious attack on Egypt had been rejected in favour of a pointless advance on Jerusalem. His war council had voted for this option, but left all the practicalities for him to solve. With no fleet to resupply them, the crusaders would have to advance like snails, making sure that food and water supplies were secured daily before moving on. And even if Jerusalem were taken, what then? It had fallen to Saladin in the first place through lack of Christian set-tlers, and to secure it, the crusaders needed mass immigration from Europe. On the one hand, there was no sign whatever of this, and on the other, it was quite clear that, once victorious, the 'pilgrims' would clamour for a return to Europe, at which point Saladin would simply occupy the Holy City again. Richard was anyway in a race against time, for the money to sustain the crusade would run out by Easter 1192 unless fresh supplies arrived from the West, and he was starting to hear alarming rumours about the upstart behaviour of his brother John in England.

Reluctant to be forced into the pointless Jerusalem venture, Richard tried a twin-track diplomatic approach. One the one hand, he promised massive bribes to the Pisans and Genoese, hitherto sworn enemies, to make common cause and back his ideas for an amphibious attack on Egypt. On the other, he strove to see what could be achieved through parleys with Saladin and in particular set out to neutralize the vizier's overtures to Conrad. Richard negotiated with Safadin and asked him to pass on to his brother a novel proposal for a new Christian kingdom of Outremer between the River Jordan and the Mediterranean, including Jerusalem.

When Saladin predictably turned this down, Richard moved on to phase two of his talks and made the extraordinary suggestion that Christians and Saracens should partition Palestine and that Safadin should marry his sister Joan and stand forth as the honest broker and guarantor of the division. Richard was playing tit for tat: Saladin had gone behind his back to inveigle the ambitious Conrad so he for his part would turn Saladin's flank by playing on the vaulting ambitions of his brother Safadin. These proposals seem quixotic and were soon blunted by the news that Joan had indignantly refused the heathen husband chosen for her. Yet the proposals were allowed to stay on the table, while both sides stalled, meanwhile continuing the military campaign of skirmish and counter-skirmish. The Saracens were busy demolishing fortresses on the road to Jerusalem and the crusaders were rebuilding others. Nerves were stretched tight on both sides, with the Muslims specializing in 24-hour pressure, alternating day-time commando raids with nocturnal attacks, using the services of

Arab irregulars who were experts in murder, abduction and horse-stealing.

It was early November before heavy fighting began again. Inching slowly forward towards Ramla, the crusaders occupied two ruined fortresses on 31 October; Richard meanwhile indulged in the usual dangerous folly of his gallery touch by charging a party of enemy scouts near Ramla in a scrimmage where he could easily have been captured. Things turned more serious on 6 November. Just 3 kilometres (2 miles) from Yasur, at Ibn-Rak, the Templars, guarding a foraging party, ran into a large detachment of Saracens and sent back urgently for help. Richard, who was supervising the rebuilding of a fortress, sent off the Earl of Leicester and the Comte de St Pol with a company of knights, but the would-be reinforcements were themselves ambushed; it turned out the Saracens had baited an elaborate double-bluff trap to draw large numbers into an ambuscade. Two separate detachments were now in danger of annihilation and to go to the rescue might seem like throwing good men after bad, or so the Lionheart's advisers counselled, since the two forward parties seemed certain to be engulfed. But it was never Richard's way to play the percentage game. Ignoring pleas to reconsider, lest he be captured himself, he called on good and true men to follow him: 'When I sent them there and asked them to go, if they die there without me then would I never again bear the name of king.' Galloping to the rescue, Richard led a furious charge, smiting, slashing, shearing and smashing, and rapidly turned the tide. Ambroise described the king in action:

> He kicked the flanks of his horse and gave him free rein and went off, faster than a sparrow-hawk. Then he galloped in among the knights, right into the Saracen people, breaking through them with such impetus that if a thunderbolt had fallen there would have been no greater destruction of their people. He pierced the ranks and

pursued them; he turned and trapped them, hewing off hands and arms and heads. They fled like beasts. Many of them were exhausted, many killed or taken. He chased them so far, following and pursuing them, until it was time to return.

WINTER QUARTERS: A TWILIGHT PERIOD

Severely bloodied, the Saracens retreated, and a week later Saladin went into winter quarters at Jerusalem. By the beginning of December, Richard decided he had no choice but to do the same, at Ramla. The weather had broken by now, and both armies suffered from rain, snow, hailstorms, mud and slush; in addition food was soggy, the salt pork rotten or rancid, clothes soaked and arms and armour rusting; the men were dirty; everything was suffused with a miasma of dampness and sick rolls mounted daily. Strong winds, low temperatures and driving sleet, which sometimes turned campsites into quagmires, made the so-called rest period a nightmare experience.

In this twilight period each side tried to manoeuvre the other in complex negotiations. Through his envoy Humphrey de Toron, Richard told Saladin that he could not leave the Holy Land until he had got at least a partition settlement. Saladin, concerned by plummeting morale and indiscipline in his own army, tried to keep the Conrad card in play but was overruled by the other emirs on his council, who stressed that any lasting peace would have to be made with Richard; Conrad was treacherous and unreliable and could not deliver on his promises. Saladin had proof of this himself on 9 November. He put it directly to Conrad's envoy that he was tired of backstairs diplomacy where nothing substantial ever happened and that his master's credibility depended on his appearing in the field against Richard with an army; this demand was greeted with the predictable silence. When Saladin held a war council on 11 November, just before going into winter quarters, he made only token resistance when the other emirs voted in favour of talks with Richard and the partition plan. By Christmas 1191 the Saracens had actually disbanded their army and there were no plans for further campaigning until the following May.

Richard himself spent Christmas at Latrun, with advance parties thrown forward as far as Beit Nuba, only 19 kilometres (12 miles) from Jerusalem. For a while the war of probe and counter-probe continued. Richard enjoyed another triumph early in the New Year when he played the Muslim double-bluff ambuscade back on a raiding party, having lain in wait all night to carry out this coup. But with morale low and desertions mounting, the Christian commanders decided on 11 January that the moment of decision could no longer be delayed. A grand war council, not just the inner cabinet of regimental commanders, convened that day to consider the feasibility of the capture of Jerusalem. Hitherto the English, French and foreign contingents, who had the 'pilgrimage' to Jerusalem as their main aim, had outvoted the native Christians but, partly because the foreign crusaders had ignored domestic advice, not least about such issues as the weather, and been proved wrong, they were by now more disposed to heed them.

One by one the local Christian leaders, and then the masters of the Templar and Hospitaller order, opined that the taking of Jerusalem was mission impossible. Their reasons were clear and cogent. Saladin had an army inside the Holy City and there was another one outside, which meant that besiegers could be caught between two fires. Camped outside Jerusalem, the crusaders would enjoy brittle and precarious supply lines with Jaffa and the coast. And, even if against the odds the attackers succeeded in taking the city, they simply did not have enough men to garrison and hold it against counter-attack, for all the foreign crusaders would depart for Europe once they had worshipped at the holy shrines; short of the mass immigration of Christians for which they hoped, but of which there was no sign, to go on to Jerusalem would be madness. Richard asked for complete clarification on this point, and it was forthcoming. Detailed maps of Jerusalem were produced, proving that the garrison left behind would not even be able to man the walls effectively, as their circumference was so great.

Since the pro-Jerusalem party could provide no convincing arguments, Richard took the sense of the meeting and ordered a return to the coast. This caused consternation in the rank and file. Richard's prestige among the troops vanished almost overnight, with many cursing his name. 'Never since the Lord God made the world was such deep grief displayed,' noted Ambroise. Other reports said that discontent was so great that a large number of old sweats could be heard publicly wondering about the truth of a God and a Christian faith that could permit such an outcome. The more sober-sided of Richard's commanders could see that his decision was right, but the Jerusalem-at-all-costs faction stressed that the crusade was just that — a do-or-die attempt to restore the holy places; Richard's grandiose plans for an amphibious attack on Egypt to destroy Saladin's power at source was a purely secular objective and had nothing to do with the cause for which the Pope had rallied the faithful.

One immediate consequence was that the hitherto well-disciplined and integrated Christian army began to break up. The French under the Duke of Burgundy, never happy at having to accept Richard's command, were the first to lead the splinter groups; they made at great speed for the fleshpots of Jaffa and Acre. Only Henry of Champagne, Richard's nephew and reliable ally, held firm, and some of the more extreme elements among the disillusioned even went to join Conrad at Tyre. Dispiritedly the rump of the crusader army trailed back to Ascalon (reached on 21 January), where the men spent a further depressing week, beginning work on new fortifications with which the Lionheart intended to make the city impregnable. Unable to be supplied by the fleet until the beginning of February because of storms, the crusaders were further cast down by the news that the contumacious Conrad had contemptuously rejected Richard's summons to bring his own forces to Ascalon.

PHONEY WAR AND INTERNAL REBELLION

Until the end of March 1192 there were no further significant clashes between crusaders and Saracens. This period of phoney war saw both sides beset by internal problems and more concerned with faction fighting than confronting the common enemy. Morale in Saladin's army was low and the desertion level high – indeed, had the crusaders known the full extent of disillusionment among the Muslims they might even have chanced a lightning dash on Jerusalem. Quite apart from the perennial lack of support from the rest of the Islamic world – Saladin frequently made vain appeals for help both to the caliph in Baghdad and the Almohads of North Africa – the Saracen leader had severe family problems. His brother Safadin (aka al-Adil) was ambitious and rumours of a rift between him and Saladin may have prompted the Lionheart's (at first sight extraordinary) proposal of marriage to his sister Joan. The idea was to increase Safadin's prestige at the expense of Saladin, since in the Muslim councils of war the other emirs already tended to favour him and the pro-Richard party, as against Saladin and his inclination to treat with Conrad; it is significant that the rumour of this marriage alliance was said to have caused rejoicing in the Saracen camp as opposed to the tepid response among the crusaders.

Of Richard's close entente with Safadin there is no doubt. The Arab sources report: 'Safadin entrapped the credulous king with shrewdness and deceived him with smooth words, so that at last they seemed to develop a sort of mutual friendship ... They parted in amity

and good spirits as warm friends.' Later Richard made the extraordinary gesture of knighting Safadin's son, prompting accusations
among the hardline crusaders that Richard was 'soft on Islam'. But
Safadin was not Saladin's only problem. His nephew Taqui ad-Din
had recently died, and his son al-Mansur was demanding immediate
confirmation of his inheritance. Saladin, not wishing to grant huge
tracts of land to a headstrong and impulsive youth, imposed strict
conditions on the bequest, which provoked al-Mansur's rebellion.

Yet Richard was scarcely in a position to take advantage of these
setbacks to Saladin, for he was facing very much the same sort of
internal rebellion himself. He tried to entice the French back for the
great rebuilding project and secured their agreement to serve until
Easter provided that they could leave at a moment's notice if they
were dissatisfied. With little choice and on the half-a-loaf principle,
Richard agreed. Bit by bit he regained his prestige among the troops,
partly by a daring raid on Darum, which rescued 1200 Christian
prisoners, and partly by getting his own hands dirty and working in
the ranks like a common soldier. For a while there was genuine
organic solidarity in the army and one observer wrote: 'You would
have seen everyone working together, chiefs, nobles, knights, squires
and men-at-arms, passing stones and rocks from hand to hand.'

Hugh of Burgundy was sufficiently intrigued to ride down to
Ascalon but, almost predictably, he soon once more fell foul of
Richard, this time because the Lionheart would not guarantee to pay
his troops after Easter. Actually what Hugh proposed was absurdly
one-sided: Richard to pay for everything yet the French to have
instantaneous opt-outs and veto powers on every single issue. But
the haughty Duke of Burgundy chose to see Richard's refusal as an
affront to his honour and stormed back to Acre. That city had meanwhile become the cockpit for fierce conflict, often degenerating into
physical violence, between the Pisans and Genoese, supporting
respectively Guy of Lusignan and Conrad of Montferrat as the true

king of Outremer. The Pisans appealed to Richard, who made another lightning ride north with his knights, which scared Conrad and his ally the Duke of Burgundy out of Acre. He patched up the feud between Pisans and Genoese, then rode north for an interview with Conrad on the road to Tyre. At an acrimonious meeting Conrad again refused to join the army at Ascalon. Richard next convened a council of notables that formally deprived Conrad of all share in the kingdom's revenues. Hugh of Burgundy then trumped Richard's ace by sending envoys to the 700 Frenchmen still at Ascalon to remind them that they were still vassals of King Philip, and he had pledged himself to Conrad. The Frenchmen took the opportunity to put heavy navvying on the walls of Ascalon behind them and depart for the carnal pleasures of Tyre. So angry was Richard at what he saw as their treachery that he refused the French access to Acre.

TORN BETWEEN ENGLAND AND PALESTINE

Playing for time by more talks with Saladin – this time the proposal was for a divided Jerusalem with the Rock and the citadel in Muslim hands and the Christians controlling everything else – Richard sought a definitive solution for the Conrad problem. This seemed even more urgent when he returned to Ascalon and found a messenger from England waiting for him. Robert, prior of Hereford, brought gloomy news about events back home, including the overthrow of the administration Richard had left behind to govern the realm and the definitive attempt by John to usurp the throne. The Lionheart needed to get back to England quickly but he could not abandon Palestine without any achievements, for he would then appear a laughing stock. On 15 April he convened a council and told his advisers that he would soon have to depart from the Holy Land and wanted a definitive solution on who was to be king of

Outremer. He personally had always favoured Guy of Lusignan but had reluctantly come to the conclusion that only the devious and treacherous Conrad was strong enough to stand up to Saladin. Guy of Lusignan, despite his loyalty to Richard, never really lived down his bad reputation as the man who had lost the battle of Hattin; his 'exploit' at Acre, moreover, had only turned out successfully when Richard arrived to lend a hand. The English king, knowing he would be accused of duplicity, manoeuvred proceedings in the council so that the decision to adopt Conrad did not seem to be of his making, then feigned surprise. To keep Guy sweet – for the Lusignans were a powerful extended clan who could make no end of trouble for Richard in the Angevin dominions in France – he came up with an ingenious idea. When he left Cyprus, he sold it to the Templars but they found the island, full of sullen and rebellious subjects, a handful and were looking for a new buyer on whom to offload it. They still owed Richard 60,000 (£39,600) of the 100,000 marks (£66,000) for the purchase and this gave the Lionheart his opportunity. He offered Guy the chance to swap his uneasy kingship of Palestine for the throne of Cyprus, for which he would simply have to repay the Templars the 40,000 marks (£26,400) deposit; Richard wrote off the rest of the debt on the fiction of a deferred 'loan' thus giving the Lusignans the bargain of the century. Everybody was satisfied.

Richard then sent Henry of Champagne to Tyre to tell Conrad the good news. Conrad was overjoyed and accepted the offer of kingship with alacrity; it was agreed that he would be crowned in Acre within a week. Before this could happen, the most unimaginable calamity occurred. On 27 April, Conrad was assassinated by a member of the Assassins – a feared, secret Islamic terror cult that took commissions from anyone for money. Who the ultimate author of Conrad's murder was will never be known, for the possible motives are as various as the putative mastermind behind the atrocity; Conrad had dozens of enemies and had alienated dozens more, including the leader of the Assassins himself. Naturally, though, Philip Augustus and Leopold of Austria found this a propaganda coup too good to miss and spread the rumour throughout Europe that the Lionheart had been the pay-master who had hired the assassins. Since the decision to elect Conrad as king had been Richard's own and he had manipulated its implementation by the crusader council, it hardly seems plausible that he would then have had his own choice liquidated; the finger of suspicion points elsewhere. In the chaos and uncertainty following the murder, the Duke of Burgundy tried to occupy the high ground by an attempt to seize Tyre but Conrad's pregnant widow, Isabella, defied him, mindful of her husband's dying instructions that she should hand the city over to Richard.

The Lionheart cut through the uncertainty by having his protégé Henry of Champagne crowned as king, on condition he married Isabella. In canon law the marriage was highly suspect both because Isabella's original marriage to Conrad had been irregular and she was pregnant. Even Henry, always Richard's puppet, had doubts about the match, and Isabella herself was extremely reluctant, but Richard browbeat the pair of them and had them married on 5 May, barely a week after Conrad's assassination, in what even those sym-pathetic to Richard thought was unseemly haste. The cleverness of this alliance was that Henry was Philip Augustus's nephew as well as Richard's, so his elevation satisfied Hugh of Burgundy. He returned to full cooperation with Richard.

A RETURN TO CAMPAIGNING

Until the crisis over Conrad, Richard had been out on horseback every day, scouting the country between Ascalon and Jerusalem, deliberately putting himself in harm's way on patrols, skirmishes and night raids, as if this were the only way he could keep frustration and boredom at bay; in mid-April there had even been a thrilling encounter with a wild boar, when the king narrowly escaped being tusked. Now, with Burgundy and the French back in the fold, he could begin campaigning in earnest. He ordered an attack on Darum, 32 kilometres (20 miles) south of Ascalon, another coastal target that, when taken, would pinch Saladin even harder, as it commanded the maritime routes from Egypt to Syria. Too impatient to wait for the entire army to appear before Darum, Richard went ahead with his knights and rendezvoused with the fleet. Saladin rushed reinforcements to the Darum garrison, but the commander there was lacklustre and did not even try to prevent the Lionheart from unloading siege engines from the vessels at anchor offshore. With his uncanny eye for the frailties of his foes, Richard had detected a weak spot in the principal tower, one of 16 that ringed the city. A brilliantly directed sapping operation covered by a constant barrage from arrow catapults undermined the weak tower in three days, and meanwhile Saladin's relieving force failed to arrive. The garrison surrendered on 22 May, giving Richard another great triumph: when the new king Henry and Hugh of Burgundy arrived shortly afterwards, he was able to tell them that the job was already done. With a keen eye for propaganda and publicity he made a great fuss about handing the fortress over to the new monarch.

Leaving a Christian garrison in Darum, the crusaders advanced to al-Hasi (the 'Cane Brake of Starlings') on 28 May, sweeping in circles to catch all Saracens in the area in a dragnet. Saladin was not yet able to make an effective response, since many of his troops had not yet returned from winter furlough. He felt himself in crisis, since hopes for an alliance with Byzantium had recently been dashed. The emperor of Constantinople, though Christian, feared the crusaders (he was right to do so as the destruction of his city by the Fourth Crusade in 1204 proved) and wanted an alliance with Saladin if the terms were right. But the emperor made too high a bid, for the conditions he suggested to Saladin were almost exactly those the vizier had rejected when made by Richard; he reasoned, why give to one set of infidels what I have already denied to another? But the Lionheart was not really in any better case, for a fresh envoy, John of Alençon, arrived from England with further tales of John's treachery, including his recent proposal for an anti-Richard alliance with the arch-enemy Philip Augustus. He informed his council that he might have to return to England soon, but the response was not what he expected. The assembled notables resolved that they would advance on Jerusalem a second time, whether Richard stayed or went home.

Like Achilles, Richard retired to sulk and was reported as throwing himself angrily on his bed once back in Ascalon. It was not just that he had been outvoted, again, by people with no sense of military realities, but the reports he received from the crusader camp were that people were actually celebrating the decision to have another shot at Jerusalem. He was going home and they were celebrating? Finally he had second thoughts. If he did not resume command, the council would elect another leader and, without his firm hand to guide them, the entire expedition would implode. There was an outside chance he could take Jerusalem in the summer and then return to England to deal with John without loss of face. He therefore sent word to his captains that he would be staying in the Holy Land until Easter 1193 or until the fall of Jerusalem, whichever happened sooner.

THE GOAL OF JERUSALEM

The Christian army set out once more on the road to Jerusalem in optimistic and even euphoric mood, despite a continuing high level of desertion. In five days the troops accomplished a march that had taken two months to complete the year before, and on the way they successfully ambushed Saracen raiders. Early in June Richard set up his headquarters at Beit Nuba, 21 kilometres (13 miles) from Jerusalem. While the crusaders waited for Henry of Champagne to complete his round up of the inevitable malingerers and backsliders who had managed to creep back to Acre, another council was held, at which Richard took revenge for the earlier slight. After repeating his arguments about the impracticability of holding on to the Holy City because of lack of Christian manpower, he announced that he was throwing up his command for the assault on Jerusalem; he would follow the others but he would not lead. 'You will not see me leading the people in this undertaking, for it will bring me blame and disgrace. You are rash in urging me into this venture. However, if you wish to head for Jerusalem now, I will not desert you. I will be your comrade, not your leader. I will follow you but not precede.'

Perplexed by this, the crusader council elected a cabinet of 20 (five from Outremer, five French, five Hospitallers and five Templars) to decide on the new leadership. Suddenly the French found themselves in a minority. The other 15 announced that Richard's strategy for an attack on Egypt was the only viable one, especially as the Lionheart had promised to finance most of it. The French protested vociferously but in vain. Once again a retreat from Jerusalem was ordered; once more morale plummeted and Richard's name was dirt; to cap all, the French withdrew from the army and set up a separate camp.

In all this the crusaders seemed to have forgotten Saladin. But he had not forgotten them and by now had prepared his counterstroke. On 12 June a large party of Saracens attacked the 'Franks' near Beit Nuba. The fight sucked in more and more troops, mainly the French, and then, by the old dodge of a feigned flight, Saladin inveigled the Templars and Hospitallers into pursuit. When the trap was sprung, ferocious hand-to-hand fighting ensued, with the Saracens at first having the advantage, until reinforcements under the Comte de Perche and the Bishop of Salisbury swung matters the crusaders' way. Richard was absent from this encounter, since he also was turning the tables on would-be ambushers at another location, the Pool of Emmaeus. He killed 20 of the enemy, captured Saladin's herald and a vast quantity of Muslim supplies in transit; Richard himself was one of the heroes that day, alongside the Earl of Leicester and Stephen Longchamp.

But it was clear that the pressure from Saladin was increasing. On 17 June there was a large-scale attack on a Christian caravan near Ramla, mainly featuring the French. First a well-timed attack by Turkish horsemen on the rearguard caused panic, and then a second wave of attackers moved in, wielding great iron clubs. Once again the French were rescued from a tight spot, and again the ubiquitous Earl of Leicester was the prime mover. Morale in the crusader army had by this time reached rock bottom: there were constant Muslim attacks, the unpopular decision to retreat from Jerusalem, and profound irritation that King Henry, supposedly rounding up backsliders in Acre, seemed to be taking an unconscionable time about it; it was almost as if he too had decided to have a recreational break among the fleshpots. Meanwhile Saladin's spirits lifted. He learnt of the crusader decision to retreat, which relieved all pressure on Jerusalem, and of the discontent in the Christian ranks. This seemed like the perfect time to strike.

'NEVER WAS THE SULTAN MORE GRIEVED'

Saladin called up a second army, recently recruited by Safadin's half-brother Falak al-Din. Coming up from Egypt with hundreds of fresh horses and mules and a wealth of trade goods, this new force seemed likely to turn the tide. Unfortunately for the Saracens, the Lionheart's intelligence service was superb. His master spy based in Cairo, a Syrian Christian named Bernard who specialized in going in disguise as a bedouin, tipped off Richard that an immense caravan was on its way north to Jerusalem. Quick thinking was always Richard's forte and in a trice he realized that Falak al-Din's caravan was crucial to the fortunes of either side. He made rapid preparations for an attack. The Duke of Burgundy and the French were prepared to join in the venture on condition they received one-third of any spoils (a greater proportion than their numbers warranted). Richard agreed and, on the evening of 21 June, set out with 500 knights and 1000 of his best infantry. A night's forced march in the moonlight brought the crusaders to the rendezvous point at Galatie, where they were resupplied by the Christians at Ascalon, who had been alerted by the English king. Meanwhile Saladin had learnt that an enemy detachment was moving southwards and correctly divined Richard's intentions; he therefore sent a large force to reinforce the caravan. On the 23rd, as Richard's raiders were converging on al-Hasi, Falak al-Din decided to take the shortest route to Jerusalem and camped at Tel-al-Khuwialifia, 23 kilometres (14 miles) away. When Saladin's reinforcements arrived, their commander angrily remonstrated with Falak for this decision, but he was adamant he would not attempt the night march recommended, fearing that his unwieldy caravan would not be able to maintain formation in the darkness. He insisted on staying where he was.

Hearing that the caravan was within striking distance, Richard could hardly believe his luck and at first suspected a trap. But when his scouts, disguised as bedouin, brought back word that the caravan was preparing to bed down normally for the night and there was nothing suspicious in their dispositions, he sensed he had an unparalleled opportunity to destroy Saladin's entire 1192 strategy. After making sure that both men and horses were well fed and watered, for dawn would bring the first of the summer's scorching desert heat, he ordered another night march. Just after daybreak, as the caparisoned camels were being led to their starting place for the trek to Jerusalem, the Saracens found themselves overwhelmed as by a storm surge. Richard's horsemen swept in among them like dervishes. The Muslim convoy and the reinforcements were so numerous that they had camped in three separate locations, and this made the crusaders' task even easier. One of the detachments was surrounded and cut to pieces; the other two fled in disarray into the desert, pursued by the triumphant ululating Christians.

Once again Richard was in the very thick of action, wielding his sword like a sabre, chopping and cutting at the foe, and at his side his two favourites, Stephen Longchamp and the Earl of Leicester, ably assisted in the butchery. The slaughter was tremendous, for the Saracens were taken completely by surprise and panicked without making any attempt at a rearguard action. Some 1300 Muslim bodies were counted on the battlefield, but the tally of deaths was much higher, since many of the wounded crawled away into the desert to die. Some 500 hundred prisoners were taken but, even more importantly, the Saracens lost 4700 camels, thousands of mules and asses, together with their precious loads of gold, silver, arms and armour, medicine, spices, clothes, tent and ropes. It was a total catastrophe for the Saracens, the kind of defeat that the crusaders could not have ridden out had fortunes been reversed. It was all Saladin could do to keep his nerve when he heard the news. The historian Baha al-Din was in the vizier's tent when the tidings came in and reported: 'Never was the Sultan more grieved or made more anxious.'

EGYPT OR JERUSALEM?

From the Christian point of view, all their immediate money worries were over. The French, now adequately financed, once again raised a clamour for an assault on Jerusalem. But Saladin, reading the enemy's mind, had guessed this would be the upshot of the débâcle suffered by the caravan and ordered a severe scorched-earth policy on the road to the Holy City, including the poisoning of all wells and water-holes on the approach roads. When Henry of Champagne (now of course king) returned from Acre with another round up of scrimshankers and malingerers, the French again demanded to have the issue debated in full council. Richard addressed the assembled notables and pointed out the grave problem of water supplies. Since Saladin had poisoned the wells, the crusader horses would have to be watered in relays in a river far from Jerusalem, with half on duty and half drinking at any one time, it would be easy for Saladin to deal with the severely reduced Christian cavalry. The French accused Richard of defeatism and called for a vote. A 'grand jury' of 300 men was chosen, from which a long-list of a dozen was selected and finally a short-list of three. It was then agreed that the decision of these three good men and true should be final: to no one's great surprise, they opted for an Egyptian campaign rather than the perilous advance on Jerusalem. The French, having given their word to abide by this decision, instantly reneged, with the Duke of Burgundy roundly declaring that no Frenchman would ever fight in Egypt. Richard then spoke again, patiently explaining that his Egyptian strategy, and particularly his planned expedition up the Nile, would force Saladin to pull back and abandon Jerusalem, so the objective of the 'pilgrims' could be fulfilled in indirect ways as well as by direct assault.

With the crusaders again despondent and in disarray, Richard once more played for time by reopening negotiations with Saladin. He warned Saladin not to draw any false conclusions from his 'strategic retreat': 'Do not be deceived by my withdrawal. The ram backs away in order to butt.' But in truth both commanders were weary and seemed to have reached stalemate. In this context the Lionheart was pleased to receive some concrete proposals from his great adversary that seemed to hold out genuine hopes of peace. Saladin offered Richard Christian possession of the contested Church of the Resurrection and free entry for pilgrims into Jerusalem; moreover Palestine would be partitioned, with the littoral towns remaining in crusader hands and the interior resting with the Saracens. Just when Richard allowed himself to become hopeful, Saladin produced the sting in the tail: in return for these 'concessions' the English king must promise to demolish Ascalon.

This was too much for Richard, who had made Ascalon his show-piece of defensive operations and spent a fortune on it. He replied to Saladin that the entire coast from Antioch to Darum had to be in crusader hands before peace could be signed. And by retreating to the coast from Beit Nubia to Acre, Richard stressed his coastal strategy as a springboard for the assault on Egypt. Saladin viewed these events differently and construed the Christian withdrawal to the coast as a sign that they were on the run. Moreover, his spies told him that, once in Acre, the Lionheart would proceed to a siege of Beirut to save face and then, with that victory under his belt, would sail for England. Saladin was still determined to win the battle of wits with his most formidable adversary and thought he saw a way forward. His next move was totally unexpected: on 27 July he ordered a surprise attack on Jaffa.

'ALAS, KING OF ENGLAND, WHAT TOOK YOU TO ACRE?'

Having encountered no opposition on the coastal march from Jaffa to Acre, which he accomplished in six days (as opposed to the agonizingly slow progress of the previous August), Richard thought the Saracens were preparing to wind down the war. Not exercising his usual caution, he had left only a weak garrison in Jaffa, together with a large number of sick and wounded. He was not to know that Saladin had just received a shot in the arm with the arrival of major Muslim reinforcements from Aleppo and Mosul. Here at last was the chance to reverse the verdict of Arsuf. Trebuchets and mangonels were brought up and began their work of aerial devastation, while sappers got to work below. Initially dismayed by this unexpected turn of events, the defenders sent word to Richard while lamenting his absence: 'Alas, king of England, our lord and protector, what took you to Acre?' The garrison was further cast down by the cowardice of the castellan Aubrey de Reims, who initially took refuge on one of the ships at anchor in the harbour and had to be shamed into returning. But they fought bravely, contesting every step into the city, formed up in the Roman turtle formation of old; only heavy stones fired from the mangonels finally cracked their resistance. All Christian fighting men then retired in good order into the citadel; by their valour they had bought a precious two days. The Saracens meanwhile spread through the city, looting, pillaging and raping and searching for prisoners to ransom. Saladin had no chance of restoring order for, as Baha al-Din reported: 'It was a long time since our troops had taken any booty or won any advantage over the enemy; they were therefore eager to take the citadel by storm.' Saladin, though, wanted to avoid further casualties and hoped he could talk the defenders in the citadel out of their eyrie.

The newly elected patriarch of Jerusalem came forward as the chief negotiator on the crusader side and proposed to Saladin that a truce be agreed, to expire at 3 p.m. next day (1 August), at which time, if no help had arrived from Acre, they would surrender and pay the vizier a large fine for his indulgence in having allowed the truce. Confident that Richard could not arrive in time to help the beleaguered garrison and pleased that he would then have taken the citadel without any loss of Saracen life, Saladin agreed. Meanwhile in Acre on 28 July the Lionheart received the desperate plea for help from the garrison: 'the life and death of all depended on him alone'. He summoned his council and asked leave to march south with the entire army, but the French refused. All other contingents being prepared to answer the call to arms, Richard divided his forces. The main army under Henry of Champagne was detailed to travel south by the overland route with the Templars and Hospitallers; this body never got into the fight, as it was halted in its tracks at Caesarea by the report that a second Muslim army stood astride the road ahead. For his strike force of marines Richard handpicked the cream of English and Angevin knights plus the crack Pisan and Genoese infantry, but luck was not with him for, standing away from Acre, the fleet was delayed by contrary winds off Mount Carmel and, in a light gale, some of the vessels were blown off course. Striding up and down the poop deck, he cursed the delay caused by the elements and was in a state of high nervous agitation until dawn on 1 August. Arriving off Jaffa with just seven ships, he thought at first he had come too late, for the shore was lined by Saracens firing arrows. The fainthearts had their usual reasons for turning back, telling the king that by now nobody was left alive in Jaffa to rescue. But, in an extraordinary feat of valour, a priest in the citadel dived from the battlements and swam through shark-infested seas to the fleet, where he explained the situation. There seemed no longer a question of a truce, for a few timid souls had left the citadel early, paid their fines to the Saracens and then been slaughtered for their pains. Those remaining behind had taken the hint and were resolved to die in a last stand rather than trust such a treacherous foe.

Richard exhorted his men and told them that a curse awaited anyone who showed less than full valour that day. He led the way overboard: 'with no armour on his legs he threw himself into the sea first, up to his groin, and forced his way powerfully on to dry land'. Although only about eighty knights landed alongside the infantry, their prowess and the accuracy of the Genoese crossbowmen soon

secured a beach-head. Baha al-Din reported a swirl of red as the crusaders advanced, partly the crimson of blood, partly the colour of the
Lionheart's waving hair and partly the deep red of his tunic and standard. Rapidly the task force fought its way up the shore to the city
and began to make inroads in the suburbs. Saladin foolishly concentrated his best efforts on getting the garrison to capitulate before the
relievers could get to them, instead of concentrating all his strength
against the maritime invaders. Once in command of the beach,
Richard ordered skirmishers to collect wood and build a palisade, to
act as a rallying point if they were driven back from the city. Then he
ordered an advance into the city, which gained ground rapidly, with
the enemy in disarray, and many reportedly still determinedly looting instead of turning round to face the new threat.

When the Saracens did finally form up in good defensive order to
confront Richard, the defenders in the garrison sortied and took
them in the rear. Soon Saladin's men were streaming away in disarray, some retreating so fast as to suggest rout, all the time pursued by
the ravenous crossbowmen, eager to add to their tally of killings.
Saladin was once more in despair. Richard was also to the fore once
more: 'The king pursued them as they ran out of the town, smashing them as a strong wind smashes ships. He believed that he should
press on in pursuit of success, so that he could not be accused of
sparing the enemies of Christ's cross.' The Saracens bolted out of Jaffa
so fast that they did not take with them any of the items from
the defeated caravan that they had temporarily retrieved and that, for
Saladin, were one of the main objectives of the capture of Jaffa.

Both sides needed a breathing space after such a lightning campaign, so the inevitable peace talks began again. Once again it was
Ascalon that was the stumbling block, with Saladin refusing to make
peace without regaining it and Richard adamant that he would neither surrender it nor demolish its defences. Once again the Lionheart
met Safadin and once again relations were cordial, though the

entente between officers and gentlemen was not matched by similar sentiments among the Christian rank and file, some of whom made a point of throwing out into a pit the bodies of the Saracen slain mixed up with the carcasses of slaughtered pigs. Saladin was confident he held all the cards in the negotiations for, as he saw it, time was on his side: Richard would have to leave soon for England or find his brother John crowned king in his absence whereas he could wage war for another decade if that was what it took.

Richard artfully stalled the talks long enough for Henry of Champagne and part of his force to arrive. Crusader manpower now numbered over 2000 – a respectable quantity but not nearly enough should Saladin launch all his might at Jaffa. To reduce the odds Richard ordered the walls of Jaffa repaired with all speed. For three days every able-bodied man in Jaffa worked night and day on the walls, the two kings (Richard and Henry) labouring alongside the ordinary troops. Yet Saladin's first move was not an all-out attack but a commando raid. Hearing that Richard and the crusader élite were camped outside the city rather than in more protected locales, Saladin sent a snatch squad to seize Richard while he slept; there would be a diversionary attack to mask the real purpose. The Saracens managed only a spectacular bungle: most of the night was spent in fervid argument between the Turkish and Mamluk irregulars as to which should act as infantry and which cavalry. It was dawn by the time the commandos finally made their bid, but their approach was spotted by a Genoese mercenary who was answering a call of nature. Aroused in time, Richard ordered his men to stand to.

'DUMBFOUNDED BY THEIR STEADFASTNESS'

Richard's instincts were correct. Furious at the fiasco of the kidnapping bid, Saladin at once ordered an all-out attack in compensation.

There was some confusion in the Christian camp, with only ten knights at first able to find horses, but the infantry formed up in good order and were further heartened when Richard made one of his rousing pre-battle speeches. Riding up and down the ranks, the English king 'urged them to be steadfast, condemning as unworthy of their race those whose spirits weakened from fear or cowardice ... True men should either triumph courageously or die gloriously.' So determined was the appearance of the Christian infantry that the Saracens seemed reluctant to attack. Baha al-Din reported: 'Like dogs of war they [the crusaders] snarled, willing to fight to the death. Our troops were frightened of them, dumbfounded by their steadfastness.'

Saladin was enraged that his men seemed fearful even when they vastly outnumbered the enemy and personally went round the divisions, trying to shore up their courage. But the Saracens had reasons for their diffidence, for Richard's dispositions were brilliant. In the front rank were kneeling men, each protected by his shield, with the butt of the lance on the ground and the shaft facing upwards like a sharpened stake. Behind this close-packed phalanx were archers, each with a bow positioned between the heads of the kneeling infantry. The archers worked in pairs, one loading and the other shooting, exchanging bows all the time so that there was continuous fire. Though heavily outnumbered, this human armadillo broke up charge after charge from the Muslim cavalry. Every time Richard sensed a faltering in the pulse of the enemy onslaught, he led a charge of his knights out from the protective cover of the infantry. All the Lionheart's legendary physical robustness was called on that day, since he fought so hard that 'the skin of his right hand tore because it was so damaged by the continual effect of brandishing his sword'. Among his valorous acts were the saving of Ralph de Mauleon from imminent capture by the Saracens and the rescue of the Earl of Leicester when he was thrown from his horse. Even more sensational an exploit was Richard's galloping along the length of his troops taunting the enemy to send an individual champion to fight him. When Saladin saw that there no takers from any of his men, he turned his back on his own army and trembled with rage. Safadin rubbed salt in the vizier's wounds by sending Richard two Arab horses in recognition of his courage even while the battle raged.

THE LIONHEART'S FINEST HOUR

All day long the noise of battle rolled, but after eight hours the Saracens had finally had enough. A final charge from Richard and his knights sent them streaming from the field. Even in a military career full of superlatives, this was the Lionheart's finest hour. Throughout the day the issue was on a knife-edge, but the king's energy, acumen and bravery won the day. At one point he was completely surrounded and seemed certain to be captured but fought so ferociously that the Saracen ranks finally parted and gave him a wide berth; he emerged from the fray covered in arrows. After Jaffa even the Saracens concluded that he was no ordinary man but rather a creature of legend.

Saladin was a fine general but every time he came up against Richard he was worsted, whether at the siege of Acre, the pitched battle of Arsuf, the attack on the caravan at Tel el-Hesi or the two battles of Jaffa. This time he had lost 700 men dead and 1500 horses, while the crusader dead barely numbered double figures (though hundreds were wounded). Baha al-Din wondered at the conundrum whereby a man who was always in the thick of the fight and seemed to be everywhere at once could still manage to direct operations with an eagle eye; Richard seemed more sorcerer than soldier. Yet the Lionheart's problems were by no means over. Saladin sent word that he would join battle again and would not cease until he had captured

'Melek Rik'. Still fearing being outnumbered, Richard sent word to the rest of Henry of Champagne's army at Caesarea to join him at Jaffa, but the men mutinied and refused to obey the order. Even worse, he himself fell ill from the exertions of the battle and wished to return to Acre to convalesce. But both at Jaffa and Ascalon those deputed to defend these fortresses in Richard's absence refused the commission, and it became clear that if he departed for Acre the garrisons would throw in the towel. Forced to stay in Jaffa, Richard was openly exasperated and distressed, raging to his intimates that he could find no one to agree with any of his plans or obey any of his wishes.

The only way forward was the usual stalling tactic of peace talks. Seriously ill, Richard tried to bluff his way out of his predicament. He sent a message to Saladin as follows: 'How long am I to make advances to the Sultan that he will not accept? More than anything I used to be keen to return to my own country, but now that winter is here and the rain has begun, I have decided to remain.' Yet Saladin was well informed both about the mutiny at Caesarea, the high desertion levels at Jaffa and the generally low morale in the Christian army. The last straw was the categorical refusal of the French to take any further part in the campaign – a snub that seemed to the superstitious to call down divine anger, for suddenly Hugh, Duke of Burgundy died.

It was in this final period in the Holy Land that the famous chivalric relations between Richard and Saladin manifested themselves. Learning that Saladin knew about his illness, Richard tried to make a virtue of his infirmity by calmly asking the Saracen leader to send him a gift of peaches and pears. Saladin prepared a basket of choice fruit and sent it together with a quantity of snow (whose origin remains mysterious) so that he could concoct an improvised fruit salad. Having been bitten many times, he was supremely shy of concluding that the crusaders were in terminal disarray, despite the encouraging reports but, by the end of August, he thought he would test the waters himself and ordered his army to advance. It was at this juncture that a new offer arrived from the Lionheart. Hitherto Ascalon had been his sticking point, for if he went home without either securing the entire coastline or taking Jerusalem, his many enemies and critics would say that his so-called feats on the Third Crusade were actually a fiasco. But now he faced the critical realization that his army might refuse to fight again, so low was morale and so high the desertion rate. Some said that by this time he was feverish and hallucinating and no longer *compos mentis*. Whatever the truth, he suddenly told Saladin that Ascalon was negotiable after all; not only that, but he was prepared to give it up without any compensation in return for a lasting peace.

RICHARD AND SALADIN: THE END-GAME

The truce was agreed on 2 September 1191. The terms were reasonable, though peace on this basis could have been obtained months before, certainly without the two bloody battles of Jaffa. Ascalon was to be demolished and not rebuilt for three years, at which time it would become the possession of whoever was living in it then; Jaffa and the coastal strip up to Acre would remain in Christian hands, and Christians would be granted safe passage through the Holy Land, free commerce and free access to Jerusalem. These terms were really a recognition of necessity for, without French cooperation, Richard could not attack Egypt as he wished and, without mass immigration from the West, there was no point in trying to take Jerusalem. What, then, could the rational objectives of any crusader army in the Holy Land be? The answer of course was that there was none, but this did not stop Richard's enemies claiming that he had sold out to Saladin, that he had given away in peace what he and his allies had won in war, and that he had demonstrated the sheer pointlessness of his 16-month campaign in Palestine. Also in Richard's mind was the pressing need to return to England and deal with John before his reprobate brother had himself crowned king.

Alive to the many criticisms of his behaviour, Richard sought to save face by claiming that the truce with Saladin was meant to last only three years, during which he would cast both John and Philip Augustus in the dust and then return to the Holy Land to resume the war with the Saracens. Saladin gracefully accepted this sporting challenge and wrote back: 'He entertained such an exalted opinion of King Richard's honour, magnanimity and general excellence, that he would rather lose his dominions to him than to any king he had ever seen – always supposing that he was obliged to lose his dominions at all.'

The truce was confirmed, and the rival leaders departed, Saladin for Jerusalem, Richard for Acre. Although some of the crusaders took the opportunity to travel to the Holy City and complete their pilgrimage, albeit not in martial triumph, Richard did not. His continuing illness may be the simplest explanation but some have speculated that he intended to lead a Fourth Crusade. If, then, he went up to Jerusalem and every last one of his troops followed him, how could he ever raise a European army in future, at the time of his second coming? For this very reason the shrewd Saladin refused Richard's suggestion of a quota system for Christian pilgrims, that only those he personally approved should be allowed to visit the sacred places. Saladin realized that the more Christian pilgrims he allowed to visit Jerusalem peacefully, the less men the Lionheart would be able to raise on any subsequent crusade.

But Richard was not able to set sail immediately as, before formal ratification, Henry of Champagne insisted that all the Saracen emirs, not just Saladin, should put their names to a bond pledging themselves to peaceful coexistence. It was thus 9 October, very late in the season for medieval Mediterranean sailing, that Richard finally left Acre on his homeward journey. Lovers of romance have always lamented that Richard and Saladin never actually met – no more did Elizabeth I and Mary Queen of Scots or General Gordon and the Mahdi – but perhaps Saladin was reluctant as he feared the sinking of his own star. Almost as if by pre-established harmony Saladin died within five months of Richard's departure (4 March 1193). The conclusion that his real life's work had been the clash with Richard is irresistible.

Richard himself survived another six years. Shipwrecked in the Adriatic and forced to return to England overland through the territories of his enemies, he adopted an incognito but was recognized and arrested just outside Vienna. That put him in the power of his old enemy Leopold of Austria, who in effect sold him on to his feudal overlord Emperor Henry VI of Germany. The German emperor, in defiance of all canons of international law (admittedly then primitive) and the crusader code, held him for ransom and demanded the enormous sum of 100,000 marks (equivalent to about £2 billion today). Amazingly, even though John intrigued with Philip Augustus to have Richard permanently confined in Germany, England raised this crippling sum, and the Lionheart was released.

Richard returned to England in triumph in March 1194. Almost exactly five years later, following a further protracted period of warfare against Philip Augustus in France, the Lionheart was killed after being wounded by a crossbow bolt near Limoges. His determination always to be in the thick of things and always to do his own reconnaissance finally proved his undoing. Venturing too close to the battlements of an enemy castle, he sustained a wound that should not have been fatal but that, after botched operations by ham-fisted surgeons, resulted in septicaemia and gangrene. The English throne passed to his brother John who, in 16 years of non-stop chaos, lost a good chunk of the Angevin empire, alienated the lands of Ireland,

Wales and Scotland, was excommunicated by the Pope and finally opposed by his barons in a civil war that only ended after his death. Richard remains England's greatest warrior–king and one of the finest captains of the Middle Ages in the West.

CHAPTER 4

Cortés

The renegade Conquistador

HERNANDO CORTÉS'S EXPLOITS were so incredible and his person-ality so magnetic that it is a surprise to find him as a late starter in the world of derring-do; most explorers and adventurers make their mark very early in life. He was born in 1484 in the town of Medellin in the province of Extremadura, where western Spain has its boundary with Portugal. The fact of being an Extremeño (as the people of Extremadura are called) was of significance for two rea-sons. Firstly, he absorbed violence almost with his mother's milk. Located in the wildest part of Castile, Extremadura was a hotbed of the stasis or strife that racked the kingdom of Castile in the late fif-teenth century – a sustained struggle for land and castles often waged by people within the same extended family – which spilled over into outright civil war in the 1470s. Secondly, Extremadura turned out to be the cradle of the conquistadores, producing (apart from Cortés) such renowned figures as Orellana, de Soto, Balboa, Valdivia and the Pizarro brothers. The eruption of poor Extremeño warriors into the Americas after 1492 can be likened to the similar explosion of displaced sailors and soldiers into African exploration at the end of the Napoleonic wars in 1815. Additionally, Cortés came from a dysfunctional family, in that his father, Martin, was 'distressed gentlefolk': an hidalgo or nobleman by birth, he made his living as a soldier of fortune. Finally, as if to signify that he were a person on whom fortune would always smile, the infant Cortés survived a near-death experience thanks to the quick thinking of his wet nurse.

A passage to the New World

Taught soldiering by his father (he became an excellent horseman), the young Hernando Cortés moved from Medellin to Salamanca in 1496, lived with his aunt and studied Latin and grammar with a view to being a lawyer. But the hard slog of legal study did not appeal to Cortés, and it is significant that the great loves of his youth were always gambling and the profession of arms. Returning to his family in Medellin at the age of 17, he found them less than happy with his cavalier way with study, so determined to try his luck in the newly discovered Caribbean islands (or 'the Indies', as they were then termed). In Seville, while waiting to take ship, he evinced the first sign of a womanizing tendency that would later become notorious: he was badly injured while climbing into – or out of – a girl's window. After suffering a bout of quartan ague or malaria in Seville, he left for Valencia, spent time as a wanderer on the road (he was said to have lived for short periods in Granada and Valladolid) until he finally managed to secure passage for the New World when he was 22; his family contacts had secured him a post as a notary in the Indies, for which a law degree was desirable but not essential.

Already Cortés's rovings had opened his eyes to a world beyond the imaginings of a Medellin boy, and the process was reinforced in the Indies. Arriving at the island of Santo Domingo with an overt get-rich-quick attitude, he was lucky enough to be taken under the wing of the governor, Nicolas de Ovando, another Extremeño. For five years he worked in relative obscurity as a notary in the town of Azua de Compostela in the west of Santo Domingo. Then opportunity beckoned: another powerful protector appeared in the form of Diego Velázquez de Cuéllar, a Spanish adventurer who conquered Cuba in 1511 and was made governor there. As Velázquez's protégé, Cortés soon showed two salient sides of his character: he became consumed with gold fever and he sired a daughter on an Indian girl.

By 1514 Cortés was already rich. He struck gold, bought a hacienda and became one of the leading lights of Cuban society. But his relationship with his patron was already following the zigzag pattern that would eventually lead Velázquez to detest him more cordially than any other human being. First there was a row over the treatment of local Indians. Velázquez was a Catholic paternalist, but Cortés wanted greater control over his indigenous peons, reducing them in effect to a form of serfdom. So bitter did the dispute become that Velázquez at first ordered his arrest but, a weak and indulgent man, he soon pardoned his protégé, as he was to do on so many occasions. There was another passage of arms in 1515 when Cortés seduced an upper-class girl named Catalina Suarez on a promise of marriage and then reneged. Once again Velázquez jailed the contumacious Cortés, but once again Cortés's glib tongue secured himself a pardon. As 'penance' he accompanied Velázquez on an expedition to put down a rebellion in northern Cuba and then cemented his restoration to favour by finally marrying Catalina, with Velázquez as witness. The marriage was unhappy from the start, with Cortés continuing to womanize and Catalina responding with permanent hypochondria; tired of her constant illness, Cortés reproached her with laziness. At all events, there were no children from the marriage. Meanwhile Cortés continued to make money from his gold mines. Flattering and cajoling Velázquez, making himself out as the prodigal son returned, he became the *alcalde* (chief magistrate) of Santiago de Cuba, then the island's largest town.

Diego Velázquez was a singularly hopeless judge of human beings, and Cortés evidently ran rings around him. The poor benighted governor regarded Cortés as his most intimate adviser and, in political terms, wholly his creature. In time Velázquez learnt to tolerate his protégé's compulsive womanizing, if only because he seemed otherwise a devout Catholic who went to Mass regularly. Cortés meanwhile bade his time, waiting for some great opportunity, never showing his political hand or revealing his true feelings to anybody. A crafty calculator, he had learnt to control his temper and smile even while he plotted treachery. His quick mind enabled him to see all the ways the law could be manipulated by a cunning lawyer and, as one of the few Latinists in Cuba, he was not slow to take advantage.

THE YEAR OF OPPORTUNITY

But at this stage no one would have marked out Cortés for greatness. Small of stature at 1.6 metres (5 feet 4 inches), he had a small head, a deep chest, was thin, somewhat bent and bow-legged and was pale-faced and brown-haired with some red tints. The year 1518 was the moment of opportunity for which he had always yearned. Velázquez had sent an expedition to colonize the Yucatán peninsula on the mainland, but the venture, commanded by his nephew, Juan de Grivalja, failed badly. Even before Grivalja limped back to Cuba, Velázquez had decided to send another expedition, this time headed by Cortés. The governor told his 'intimate friend' that he would supply and pay for two ships but Cortés must find any others he wished to take. Cortés was ordered to seize all new lands in the name of the king of Spain, to discover and populate new territories in the service of God and to share any profits from the exploration with his patron. Perhaps knowing Cortés's sexual nature at least, if not his deeper thoughts, Velázquez included in his written instructions the somewhat curious proviso that the Spanish were to abstain from sexual intercourse with the native women.

Cortés could scarcely believe his good fortune. Once clear of Cuba he intended to be his own man and to shake off the island's dust and Velázquez's tutelage for ever. He made elaborate preparations in the port of Santiago de Cuba, adding more and more ships and men to the expedition. As his second in command he recruited Pedro de Alvarado, a veteran of the Grivalja probe into Yucatán, and began to boast about the wealth they would find in the new territories; Alvarado seemed the perfect foil, his impetuous openness appealing to the circumspect, devious Cortés. Finally alerted by spies about Cortés's disloyal talk and the scale of his preparations, Velázquez, typically, dithered too long before moving decisively against his so-called friend. By the time he decided to replace Cortés as leader of the expedition or possibly even halt it altogether, Cortés was already on his way.

And now for the first time Cortés revealed a murderous quality that would characterize his career ever afterwards. Velázquez sent a messenger to the port, replacing Cortés as leader with a man named Luis de Medina. The messenger was stabbed to death by Cortés's accomplice, Juan Suarez, and the governor's papers taken to Cortés. Realizing the way the wind was blowing, Cortés quickly bought up all the meat in Santiago de Cuba and fully provisioned his six ships, intending to present the governor with a *fait accompli*. At the very last moment, Velázquez raced down to Santiago de Cuba in person just as the fleet was setting sail. According to one story, Velázquez yelled reproaches from the shore at the departing Cortés, to which he replied with bland, hypocritical reassurances. Needless to say, he did not put back to shore.

CORTÉS

'THE REBEL CORTÉS'

Departing on 18 November 1518, Cortés cruised along the southern coast of Cuba, making further landfalls and recruiting further adventurers there. For the first time his powers of verbal persuasion, the gift of the gab, were fully on view as he persuaded official after official to do his will, even though in some cases they already had express orders from the governor not to cooperate with 'the rebel Cortés'. He did well in this regard at Trinidad, but at San Cristóbal de Havana (the embryo of the later city of Havana but at that time situated on the southwest coast) he found most of the inhabitants staunchly loyal to Velázquez. It was too little, too late, for by this time the governor, never a strong or decisive man, had virtually abandoned hope of stopping his one-time protégé. He contented himself with loud complaints about Cortés's perfidy, portraying him as an egregious ingrate who had taken his (Velázquez's) money and run; actually the probable cost of the expedition was about 60–40 in Velázquez's favour. Finally, at Cape Corrientes, at the extreme western tip of Cuba, Cortés held a muster that revealed that his expedition contained 530 Europeans, of whom 30 were armed with crossbows and 12 had harquebuses; on board also were 14 pieces of light artillery, 16 horses and a large number of mastiffs – for the Spanish were accustomed to enter battle with huge fighting dogs. The adventurers had enough bread, bacon, oil and wine for a two-week voyage but barely enough water for the crossing to the Yucatán peninsula.

Cortés hoist sail for Yucatán on 18 February 1519. After making landfall he began cruising northward along the coast, putting in to shore deliberately to test the mettle of the Mayas (heirs to the great Maya civilization that had been at its peak in the eleventh century), who had worsted Grivalja's men in the first Spanish expedition to Yucatán. Cortés decisively defeated a Mayan army at Potonchán and impressed his men with the efficiency of the light artillery. In the

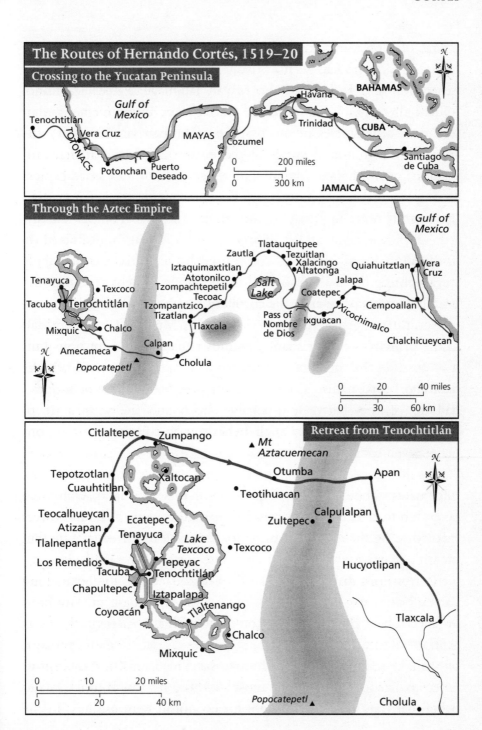

The Routes of Hernándo Cortés, 1519–20

Crossing to the Yucatan Peninsula

BAHAMAS

Gulf of Mexico

Havana

Tenochtitlán

Vera Cruz

TOTONACS

MAYAS

Cozumel

Trinidad

CUBA

Potonchan

Puerto Deseado

Santiago de Cuba

0   200 miles
0   300 km

JAMAICA

Through the Aztec Empire

Gulf of Mexico

Tlatauquitpee

Zautla

Tezuitlan

Iztaquimaxtitlan

Xalacingo

Altatonga

Atotonilco

Quiahuiztlan

Vera Cruz

Tzompachtepetil

Jalapa

Tenayuca

Texcoco

Tecoac

Salt Lake

Coatepec

Tacuba

Tenochtitlán

Tzompantzico

Cempoallan

Tizatlan

Xicochimalco

Mixquic

Chalco

Tlaxcala

Pass of Nombre de Dios

Ixguacan

Chalchicueycan

N

Amecameca

Calpan

Popocatepetl

Cholula

0   20   40 miles
0   30   60 km

Retreat from Tenochtitlán

Citlaltepec

Zumpango

Mt Aztacuemecan

Otumba

Apan

N

Tepotzotlan

Xaltocan

Cuauhtitlan

Teotihuacan

Teocalhueycan

Calpulalpan

Atizapan

Ecatepec

Zultepec

Tlalnepantla

Tenayuca

Lake Texcoco

Texcoco

Los Remedios

Tepeyac

Tacuba

Tenochtitlán

Hucyotlipan

Chapultepec

Iztapalapa

Coyoacán

Tlaltenango

Tlaxcala

Chalco

Mixquic

0   10   20 miles
0   20   40 km

Popocatepetl

Cholula

181

course of these early skirmishes with the Maya, Cortés managed to rescue a man who had been a prisoner of the Maya since the Grivalja expedition and had learnt the Mayan language; Father Jeronimo Aguilar thereafter became one of Cortés's key interpreters. Aguilar told gruesome stories of the practice of human sacrifice that was apparently common throughout the entire length and breadth of what is now Mexico. The incontestable fact of human sacrifice among the Mayas and Aztecs gave Cortés what he needed: a crucial propaganda point in justifying the cruelty and barbarism he habitually employed against the native peoples. Cortés's overarching justification for all atrocities committed by his men was that they were fighting the truly diabolical in the shape of people who practised human sacrifice and, as a corollary, that he was fighting hard to convert the indigenous tribes to Christianity. Much of Cortés's appeal to the principles of Christianity sounds ritualistic and spurious but it is possible that genuine devoutness coexisted in his soul with a dark and brutal nature. Cortés, while a true believer, seems to have had no conflict between ostensible creed and the most wicked behaviour; one might say such behaviour is common in religion beliefs even today.

The battle of Potonchán gave Cortés another crucial piece of information about the native peoples of this new land. Outnumbered ten to one, the Spanish had nonetheless won easily, and there seemed to be three main reasons for this. The terror caused by the unfamiliar artillery was one, and the inferiority of the Mayan swords made from obsidian (sharp, black volcanic rock) to the Spaniards' Toledo blades another. But, most of all, the Mayan inferiority in battle was caused by culture. In common with the Aztecs and other central American tribes they entered battle not to kill their enemies but to take them prisoner for human sacrifice later. The Mayan and Aztec civilizations provide the most stunning refutation of the facile belief that wars are always fought for economic reasons.

Cortés thus had three trump cards and discovered a fourth in his next large-scale battle with the Maya. Towards the end of March he defeated them at Centla. Having planned a flank attack, he and his other horsemen were delayed by swamps and then appeared on the field of battle almost miraculously at just the right time. At sight of the horses, unknown in the Americas, the Mayans panicked, thinking that horse and rider were one creature. However, it is absurd to suggest that horses played any major part in the eventual conquest of Mexico. The Indian tribes soon got used to them and feared them no more once they discovered they were mortal. In general, the effect of superior Spanish weaponry, horses and Indian superstition has been absurdly overplayed in traditional accounts of Cortés. But there was one dramatic result of the battle of Centla. The defeated Mayans presented Cortés with a number of women and among them was one who came to be known as Doña Marina or La Malinche. It seems that her real name was Malinali, and she was supposedly the daughter of a local Indian lord who had been sold by her parents and then enslaved by the Mayans. Marina spoke the Aztec language Nahuatl as her mother tongue but also knew Mayan from her years in captivity. Highly intelligent, she proved invaluable to Cortés, for she could translate Nahuatl into Mayan for Aguilar who then translated into Spanish. Though not especially beautiful, she became indispensable to her new master who made her his mistress and added her to the informal harem he had begun to assemble. Her loyalty to Cortés was absolute and eventually she bore him a son, Martin.

In Aztec lands

After christening Marina and 'converting' the Maya by requiring their attendance at Mass, Cortés and his men set sail on Palm Sunday 1519 and anchored at San Juan de Ulúa on Maundy Thursday. They were now in the territory of the Aztecs, and it was just two days later that the first emissaries from their emperor Montezuma arrived. When Cortés arrived in his land, Montezuma was aged 50, having come to the throne in 1502. The Aztecs were the dominant power in what is now Mexico (indeed they called themselves the Mexica) and had established their hegemony over the other tribes in the fifteenth century, after beginning their drive to power around 1430, after which date they waged a war of continuous conquest. Montezuma, fifth in the line of Aztec emperors and the eighth ruler since his dynasty was founded at the end of the fourteenth century, lived in splendour in Tenochtitlán, on the site of what is now Mexico City. Situated at an altitude of just over 2000 metres (7000 feet) on an island near the shore of a great lake, the city enjoyed spectacular views around the broad valley running down to the lake and the high mountains beyond (two of them volcanoes).

The Aztecs had turned Tenochtitlán into a New World version of Venice by driving stakes into the bed of the lake, building causeways around some 1000 hectares (2500 acres), constructing three bridges linking the 'impregnable' city with the mainland, and filling up any gaps between the causeways and bridges with mud and stones. All around the lake were tribute-bearing towns and villages making obeisance to the great city in their middle; the lake itself positively teemed with canoes of all kinds. With 250,000 inhabitants, Tenochtitlán was larger than any city in Europe except Naples and Constantinople. In the centre of the city was a walled holy precinct, with numerous sacred buildings and pyramids topped by temples. The core of the Aztec power nexus was its alliance with the two important towns of Tacuba to the west and Texcoco on the east of the lake shore. Within the obvious limitations of the detestable practice of human sacrifice – which the Aztecs had certainly not invented, though they increased its use for religious and social reasons – Montezuma presided over a stable society, ossified in its hierarchies and hidebound in its way of doing things. Crime was rare, and the Aztec culture was, in its own way, highly moral, with a clear differentiation between right and wrong. The principal defect of the culture was a pessimistic world-view, for Aztec myth looked forward to an Armageddon, very like that of Ragnarok in the Norse myths, when all would be destroyed and collapse into darkness.

What seemed to be a mighty empire had an Achilles' heel, or rather two of them. Although Montezuma's writ ran right down to the Caribbean coastline (Cortés had landed near modern Vera Cruz), the tribes between that coast and Tenochtitlán suffered the Aztec yoke in brooding, sullen silence, alert for any chance that might allow them to subvert their conquerors. Particularly volatile in this regard were the Tlaxcalans, who paid tribute under duress but nursed a bitter hatred of the Mexica people. To make matters worse for the Aztecs, Tenochtitlán was overpopulated and could not feed itself; it depended heavily on tribute from subject peoples. By the time of Montezuma, Aztec society had compounded the problem by becoming more and more stratified, with rigid social distinctions. The upper classes had developed a taste for luxury goods, which meant ever more tribute to pay for them, thus increasing the brittleness of their society. Montezuma had dealt with all signs of disaffection both within Tenochtitlán and in the wider empire with utmost ruthlessness, but when his spies brought him word of the coming of mysterious white men he was nonplussed. Who were these people who had arrived from the sea in 'towers or small mountains'? Montezuma held long conferences with his magicians, priests and shamans as well as his cabinet of advisers. There has been a long and persistent myth to the effect that the Aztecs regarded the conquistadores as supermen and Cortés as a returning god, perhaps Queztlcoatl. The idea of the Spaniards as gods is a later invention, a myth spun to the Indians by Christian friars to help legitimate the rule of the king of Spain. It is clear that most of Montezuma's

185

advisers viewed the newcomers as criminals or terrorists. We cannot follow all the agonized deliberations of the Aztecs, except that they sent embassy after embassy down to the coast to try to learn more about the strangers. There must have been counsellors who advised Montezuma to act decisively, to show no mercy and kill the 'wizards'. For whatever reason, the emperor decided instead on a policy of appeasement.

A MASTER-MACHIAVELLIAN

This handed the psychological advantage to Cortés, for such a man viewed diplomacy and conciliation simply as weakness. An expert in mind games, he began by systematically browbeating and terrifying Montezuma's envoys, impressing them with his horses and guns. Soon he asked what for him was the vital question: did Montezuma possess gold? An expert liar, Cortés demonstrated the veracity of the old adage – the best way of telling a lie is to tell the truth – by informing the Aztec emissaries that he had a disease of the heart that could be cured only by gold – true enough, though scarcely the manifest meaning of his remark. The principal Aztec envoy Teudile informed Montezuma of this, and the emperor sent back a helmet full of gold, naively imagining that this would satisfy Cortés and he would depart; naturally, it merely whetted his appetite. Meanwhile, the conferences at Tenochtitlán continued. A majority of the emperor's advisers, including the later ruler Cuitlahuac, were adamant that the Spanish should not be allowed to proceed further into their kingdom, but Montezuma prevaricated. He decided to stall by inventing myriad reasons why the Spanish could not travel up to Tenochtitlán nor he to the coast. But he was wasting valuable time and allowing the initiative to pass to Cortés. Already Cortés had told his intimate friends that the evidence of widespread human sacrifice had made him resolved to be merciless in his dealings with the Aztecs. A signal turning point in his fortunes occurred when he was visited by some coastal Indians of the Totonac tribe and for the first time realized how deeply hated the Aztecs were by their subject peoples. It was now that he first began to crystallize his ideas for fomenting an anti-Mexica alliance among the other Indian tribes. He knew only too well that the famous conquest of Granada, which Ferdinand and Isabella had completed in 1492, had been greatly facilitated by divisions among the Moors.

But now appeared another problem – one that was to beset Cortés for the next two years. Among his men there was a significant minority who, their lust for gold notwithstanding, were primarily loyal to Velázquez, the governor of Cuba. Increasingly this pro-Velázquez faction argued for either an immediate return to Cuba or the founding of a settlement on the coast. Cortés identified his chief opponents and played his favourite game of divide and rule by dispatching them on various so-called important diversionary expeditions. While these men were away, he staged a virtual coup d'état by having himself declared Captain-General of New Spain (as Mexico was originally called). Cortés announced that his loyalty to the Spanish crown was unflinching but that, as a sovereign governor, he no longer owed allegiance to the governor of Cuba or took his orders. As a sop to the rival faction, and partly to disguise his true intentions, he founded a new town of Villa Rica de la Vera Cruz (later Vera Cruz) on 28 June 1519.

But his true machiavellianism manifested itself a little later. Hearing from freelance adventurers who arrived in Mexico in his wake that Velázquez had been granted important new powers by the Spanish crown, which obviously overrode his own self-ascribed status as Captain-General, he decide to send ships back directly to Spain, bypassing Cuba, to plead his case at Madrid. The brilliance of his stratagem was that it enabled him to kill two birds with one stone:

as his envoys to Spain he sent all the principal leaders of the pro-Velázquez faction. A master of obfuscation, logic-chopping, legalese and barrack-room lawyering, Cortés knew that bureaucracies love paperwork. He therefore sent back to Spain a veritable mountain of documents, essentially a farrago of legal mumbo-jumbo concerning the founding of Villa Rica de Vera Cruz and his steadfast loyalty to the crown. Using mystification and the cuttlefish tactics of deliberate turbidity, he sought to persuade the Spanish monarch, Charles V, to accept his de facto position in Mexico; being Cortés, he sugared the pill by sending the king all the gold he had received from the Aztecs. Velázquez's supporters tried to seize a ship so as to sail back to Cuba and warn the governor to intercept the embassy to Spain. But the plot was betrayed to Cortés, and he hanged the ringleaders, using terror to cow the opposition in his own ranks as much as he had used it to dismay the Aztecs.

To CONQUER OR TO DIE

Cortés then moved quickly to advance his designs against the Aztecs. He defeated the local Mexica garrison and told the local tribe, the Totonacs, that he was their friend and protector and that they should no longer pay tribute to Montezuma. As the last stage of his preparations to march on Tenochtitlán, he ordered all his ships beached and destroyed; all but three of the vessels were disabled. Contrary to the famous legend, he did not burn them – this is yet another Cortés myth. But it has been suggested that, always the master-machiavellian, he bowed to a 'suggestion' made by one of his followers to scuttle the ships, so that he could avoid any blame later on. He then announced to his men that Tenochtitlán was his goal. He painted a lurid picture of the booty that would be theirs and in effect proclaimed his royal status by saying that when the king of Spain had been given the 'royal fifth' (*quinta real*) of any treasure, as per custom, he himself would also take a fifth. It is not entirely clear what his true intentions were at this stage: probably he wished merely to install himself as the power behind Montezuma's throne and be a kind of grey eminence, controlling the destinies of Mexico. But to his men he declared rhetorically that his aim was to conquer or to die, that great riches awaited them but they should have no illusions about the grim obstacles that lay ahead: a 400-kilometre (250-mile) march, at first through low-lying tropics, then through a 2000-metre (6000-foot) pass into a range of temperate mountains, across another plain and finally an even more strenuous set of mountains (they would have to cross a pass at 4000 metres/13,000 feet) before debouching on the plain that lay before the lake and Tenochtitlán.

On 8 August he set out on his great endeavour. With him were 300 Castilians in six companies of 50, some 40 crossbowmen, 20 harquebusiers, 3 pieces of artillery, 5 horses and several packs of snarling mastiffs. The news of his departure sent Montezuma into gloomy inertia, but he was constrained by his own codes and culture. Since Cortés stated that he was coming on an 'embassy', by the Aztec rules of conduct Montezuma had to receive him, because (in his view) if Cortés had warlike intent, he should have issued a formal declaration of war. With his uncanny nose for weakness, Cortés had already discerned this fatal weakness in the cultural armoury of his opponent. And on he came, remorselessly. The going was every bit as tough as he had warned, with the Spanish suffering hunger and thirst in the daytime when crossing desert country and then excessive cold at night.

The initial contacts with Indian tribes were not as promising as Cortés had hoped. He asked one chief if he were Montezuma's vassal, and the cacique looked surprised: he asked who was not a vassal of the Aztec ruler, since he was king of the whole world. Even worse, the Tlaxcalans occupying the middle region between the coast and Tenochtitlán made it plain they would oppose him, whereas Cortés had counted on being able to detach them from Aztec suzerainty. When their warriors appeared plumed and fully caparisoned for battle, he quickly defeated their rearguard, but the Tlaxcalans killed two of the horses and discovered they were mortal creatures and not magical. Cortés was seriously depressed by the Tlaxcalan opposition – 'at that moment we were lean, weary and unhappy about this war, of which we could neither see nor forecast the end' – but comforted himself from some key discoveries he had made during the battle. The Tlaxcalans confirmed that they, along with the Aztec and other tribes, fought only to capture enemies, not to kill. Moreover, they fought in an absurdly wasteful way, a head-on combat, in which only the front rank engaged at any one time; when this was annihilated, the second rank moved up and so on. If this were how all Indians fought, Cortés concluded, his task would be simple, for Toledo steel was clearly superior to the obsidian swords in hand-to-hand combat, and his artillery could blast great holes in the packed mass of warriors awaiting their turn in the front line.

CYCLES OF VIOLENCE AND FORTUNE'S FAVOUR

Further good news for Cortés came when two Tlaxcalan envoys arrived with a face-saving tale that the attack had been launched by a breakaway sept of the Otomi people. But the Spanish remained cautious and for two weeks dug themselves in on a hilltop. They were right to do so, for it seems that the Tlaxcalans were playing Cortés's own machiavellian games against him. They aimed to lull the white men, then slaughter them, but they were hedging their bets with the envoys: if they destroyed the Spanish, all well and good; if not, they would simply say it was the Otomi renegades again. It later turned out that the repudiated Otomi attack was all part of this devious plan. Another Otomi ambush was attempted, to which Cortés replied with his favourite 'shock and awe' tactics, sending out his men to retaliate against the civilian population with studied atrocities. On or about 5 September the Tlaxcalans launched a surprise and determined night attack against the Spanish positions and were beaten off with difficulty.

This was the hardest-fought battle the conquistadores had experienced yet, and it continued into the second day. It was probably only the artillery that turned the tide, as the Spanish cannon balls sowed genuine terror in the enemy. Cortés again responded with studied atrocities, ratcheting up the cycle of violence another notch. When the Tlaxcalans withdrew to lick their wounds, Cortés found he had another crisis of confidence on his hands. By this time he had lost about fifty men in battle or to disease since leaving Cuba, and morale was plummeting, with a majority in favour of an immediate return to Vera Cruz. As Cortés later reported to Emperor Charles V of Spain: 'There was amongst us not one who was not very much afraid ... Many times I was asked to turn back ... I encouraged them by reminding them that they were your majesty's vassals ... moreover as Christians we were obliged to wage war against the enemies of our faith ... They should observe that God was on our side.'

Cortés's gambler's luck held. The Tlaxcalans were even more demoralized by the battles' results than the conquistadores. As one of their leaders pointed out: 'The Otomi was a brave warrior but he was helpless against them: they scorned him as a mere nothing.' The man who had most vehemently argued for war against the Spanish, a chief named Xicotencatl the Younger, went to the Spanish camp with peace proposals, saying the Tlaxcalans would obey Cortés's commands if he admitted them to his anti-Aztec alliance. Cortés knew when to leave well alone: he told the envoy that the Tlaxcalans were to blame for what had happened but he was interested only in a lasting peace. To his own men, and especially the doubters, he pointed out the brilliant success of his 'leapfrogging' tactics: he had used Totonacs successfully as allies against the Tlaxcalans and now he would use the Tlaxcalans in the same way against the Aztecs. He set the seal on his triumph by entering the city of Tlaxcala in ostentatious pomp on 18 September and enjoying a great feast that the defeated had laid out for him.

There followed 20 days of rest and recreation for the weary Spaniards. The Tlaxcalans presented the conquerors with women, whom Cortés shared out among his most meritorious soldiers. There was some attempt to explain Christianity to the bemused hosts, but a friar rightly advised the Spanish not to push religion at this stage. Cortés used his time well, learning in great detail about Tenochtitlán, the Aztec leaders and the layout of the city. He mightily impressed the Tlaxcalans with his personality, charisma and eloquence, but showed himself a master politician by insisting that his men respect the local religion and folk-ways. Both sides looked forward to the coming conflict with the Aztecs. The Tlaxcalans saw that with the help of the strangers they could at last defeat the dreaded Mexica whom they had considered invincible. For Cortés, the stay in Tlaxcala was a notable turning point. His ambitions were no longer 'fortune favours the brave' dreams, to use his favourite Latin tag, but were firmly grounded in reality. The conquest of Mexico now looked well within his grasp.

MASS SLAUGHTER AT CHOLULA

On 12 October the expedition marched down to Cholula, a great city with a population of 180,000 and 430 pyramids. The Tlaxcalans had warned him not to take this route to Tenochtitlán, on the precise grounds that it was the itinerary Montezuma suggested to Cortés through his envoys. But for the conquistadores credibility dictated that they should not shirk from this encounter. Cortés began by sending a message to the Cholulans as peremptory as it was absurd: he told them they must either submit to the king of Spain or be treated as rebels – a patent illogicality. From their point of view the Cholulans were under two constraints: Montezuma was urging them to oppose the strangers, and their own food supplies were perilously low – primitive societies usually lived on the very edge of subsistence. Yet the Cholulans were ambivalent about heeding the Aztecs' orders, and Cortés, with his super-sensitive antennae, at once picked up on this. Another of his favourite mottoes was 'a kingdom divided against itself must fall', and here he was witnessing a textbook example. As he wrote later: 'When I saw the discord and animosity between these two peoples I was not a little pleased, for it seemed to further my purpose considerably.' He decided that the way forward was via a pre-emptive strike. Since the Cholulan envoys claimed that what particularly worried them was the presence of the Tlaxcalans in the Spanish ranks, Cortés left his allies outside the city gates and went in with his conquistadores.

His suspicions were immediately aroused by the absence of people in the city and the building of barricades. What happened next cannot be reconstructed in detail since the sources are confused and contradictory. Either there was a genuine plot by the Cholulans to massacre the newcomers or Cortés persuaded himself that there was. In any event, he invited the Cholulan aristocrats to assemble in the temple courtyard for a grand conference, then sealed all exits and unleashed his warriors to massacre them; their blood up with the original slaughter, the Spanish then cut loose on a general orgy of mayhem. The Tlaxcalans and Totonac allies then rushed in and joined with gusto in the general rapine, sacking the city so thoroughly that it took them two days to exhaust their homicidal impulses. The benighted Cholulans at first thought Quetzlcoatl would protect them but, once the idols were destroyed and nothing happened, they gave themselves over to despair; many of those who had survived the original massacre hurled themselves from temple roofs. In the general madness Cortés's orders that no women and children were to be harmed counted for nothing. When they had finally slaked their bloodlust, he ordered the Tlaxacalans to swear eternal friendship to the remaining Cholulans. Montezuma's ambassadors, who were with Cortés and witnessed the massacre, were suitably horrified.

Their reports did nothing to calm the restless spirits of Montezuma in Tencohtitlan, which were further depressed by a machiavellian message from the Spanish leader. Cortés told Montezuma he knew he was at the bottom of the Cholulan plot but forgave him. Montezuma protested indignantly that he knew nothing of any plot. Cortés pretended to be satisfied with this and sent another message to say that he did not now believe that the Aztecs had plotted against him, but would soon be in Tenochtitlán to give his assurances in person. Terrified as much by this as the detailed reports of the massacre, Montezuma replied that he would be glad to entertain the Spanish but he had no food. Unfortunately by this time he had cried wolf once too often, inventing a string of excuses – health, religion, impassable roads – as to why he could not meet Cortés. The conquistadores had had the food-shortage excuse before but even in Cholula, far inferior in resources to Tencochtitlan, they found enough food to enable them to enjoy another fortnight of rest and recreation. When the expedition set out again on or about 1 November, Spanish morale was high.

An inexorable advance

Cortés was by now confident enough to allow his early allies, the Totonacs and their neighbours the Cempoalans, to depart for home

with their booty, keeping only the Tlaxcalans as allies. It was 80 kilometres (50 miles) from Cholula to Tenochtitlán as the crow flies, but
ahead loomed the high ranges with snow-covered mountain passes.
Initially the conquistadores marched through villages friendly to
Tlaxcala, but as they began the mountain ascent they came upon a
parting of the ways, with one road blocked and another open. First-
rate Tlaxcalan intelligence revealed that the Aztecs were preparing an
ambush on the open road, so Cortés simply had his men remove the
plants and trees obstructing the blocked road and proceeded on it,
leaving the ambuscade high and dry. Increasingly desperate to stop
the inexorable advance of the horrifying white men, Montezuma
tried every trick he knew to slow Cortés down, even at one point dispatching a double of himself to meet him, hoping the Spanish
would go home once they had seen him. He sent them hoards of
gold, which simply increased the Spanish desire to lay their hands
on even more. Meanwhile the majority of his council urged him to
kill the strangers. The most perceptive Aztec chief, Cuitlahuac, told
Montezuma roundly: 'I pray to our gods that you will not let the
strangers into your house. They will cast you out of it and overthrow
your rule, and when you try to recover what you have lost, it will be
too late.'

The Aztec emperor was still dithering about what to do next
when Cortés swept down from the snow-covered passes and came to
the southeast corner of the great lake at Chalco. He then proceeded
to Mixquic, to Cuitlahuac, the first town on an island, and thence
to Itxapalapa, on the far side of the Tenochtitlán causeway. On 8
November the expedition finally crossed the causeway itself, finding
it 40 metres (140 feet) wide and 'a lance's length' above the water.
The conquistadores were amazed at what spread out before them.
Bernal Diaz, the best known chronicler of Cortes's exploits, wrote
that it was 'like an enchanted vision from the tale of Amadis. Indeed,
some of our soldiers asked if it was not all a dream … It was all so

wonderful that I do not know how to describe this first glimpse of things never heard of, seen or dreamed before.'

Cortés reported to Emperor Charles V:

> The principal streets are very wide and very straight. Some of these, and all the smaller streets, are made as to one half of earth, while the other is a canal by which the Indians travel in boats, and all these streets from one end of the town to the other are opened in such a way that the water can completely cross them. All these openings – and some are very wide – are spanned by bridges made of very solid and well-worked beams, so that across many of them ten horsemen can ride abreast ... Montezuma had a place in the town of such a kind, and so marvellous, that it seems to me almost impossible to describe its beauty and magnificence. I will say no more than that we have nothing like it in Spain.

The only ominous sight was the thousands of canoes plying the waters of the lake. As Diaz said: 'We were scarcely 400 strong and we well remembered the words and warnings.' But the eventual meeting of Cortés and Montezuma passed without incident. The emperor arrived on a magnificent litter, he and Cortés exchanged formulaic greetings, and the visitors were shown to their quarters in the palace of Axayacatl.

The next few days were taken up with escorted sightseeing. Cortés and Montezuma met several times and Montezuma even made what Diaz said was a good speech, which Cortés disingenuously claimed meant he accepted to be a vassal of the king of Spain. Yet after a few days of superficial civilities the Spanish leaders grew anxious. They feared that they were flies in a spider's web, and the apprehension increased after the visit to Huichilobos, the greatest temple in Mexico. From the high vantage point on the roof Cortés's lieutenants could see how easy it would be to get trapped in the island city, with its removable bridges and destructible causeways. Further anxiety was caused when Cortés foolishly raised the issue of Christianity and asked if he could erect a cross and a statue of the Madonna in the temple. Montezuma and his priest reacted angrily to this sacrilegious talk, and Cortés was hard put to it to appease them. The frayed nerves of the conquistadores were hardly improved by Montezuma's insistence that he continue with his usual round of human sacrifices while he entertained his guests. The only bright spot for the gold-fevered conquistadores was that when building a Christian altar in their own quarters they unexpectedly uncovered a sealed doorway that led into a vault containing an immense treasure.

A RUTHLESS ABDUCTION

Cortés for a time seemed content just exploring the byways of this brave, new world. He was particularly taken with Tenochtitlán's twin city, Tlatelolco, the commercial and market centre. To Emperor Charles V he wrote:

> It is ... twice the size of the town of Salamanca, completely
> surrounded by arcades, where every day there are more than sixty
> thousand souls who buy and sell, and there are all kinds of
> merchandises from all the provinces, whether it is provisions,
> victuals or jewels of gold and silver. There is nothing to be found
> in all the land which is not sold in these markets ... Each kind of
> merchandise is sold in its own particular street, and no other kind
> may be sold there: this rule is very well enforced. All is sold by
> number and measure, but up till now no weighing by balance has
> been observed. A very fine building in the great square serves as a
> kind of audience chamber where ten or a dozen persons are always

seated, as judges, who deliberate on all cases arising in the market and pass sentence on evil doers ... I have seen them destroy measures which are false.

Markets, indeed, seemed to obsess the conquistadores. Bernal Diaz wrote that the Aztec capital outstripped Salamanca by far and marvelled at the smooth, white flagstones in the open courts, larger even than the plaza at Salamanca: 'Some of our soldiers, who had been in many parts of the world, in Constantinople, in Rome and all over Italy, said they had never seen a market so well laid out, so large, so orderly and so full of people.'

If Cortés were enjoying himself, his men were jumpy and nervous. Both 'push' and 'pull' factors were now impelling him to some decisive action. On the one hand, his lieutenants urged him to move decisively before they were all trapped like rats; on the other, they were actuated to new acts of daring by the sight of gold beyond their dreams. On 14 November Cortés carried out one of the most spectacular coups in history. He and a large body of his men went to an audience with Montezuma. Using the trumped-up excuse that the Aztecs had attacked his men on the coast (actually a greedy conquistador had provoked a fight over – inevitably – gold), Cortés told Montezuma that he must accompany him to his quarters, where he would in effect be under house arrest. 'My person is not one to be taken captive,' Montezuma boomed, and at first was incredulous at this impertinence from the strangers. Cortés told him he must come with them or be killed then and there; Marina joined in with an impassioned harangue of the emperor that convinced him the Spanish were in earnest. But the wrangling took up most of the day. In the end, to cut through the impasse, Montezuma agreed to go with his captors.

The kidnapping of Montezuma was one of the turning points in the conquest of Mexico and possibly the most important single event. What it illustrated, apart from the deadly ruthlessness of Cortés, was the superiority of the European Renaissance culture to that of the Aztec. Montezuma should have resisted and died a martyr's death, which would have inflamed the Mexica against the interlopers; they would surely never have escaped alive. But Montezuma's mind-set was blinkered: he had never been confronted by a situation like this, which to him was inconceivable, and simply lacked the imagination and mental resources to think his way out of it. Instead of taking up arms when they heard of his abduction, the Aztecs lapsed into sullen apathy and listlessness. Their culture stifled initiative, being a traditional rule-based society. There was no precedent for the kidnap of a ruler – it was not covered by the Mexica code, there was nothing about it in their book of rules – so they reacted with incredulity, stupefaction and inertia. Cortés, on the other hand, was imbued with the Renaissance ethos that boldness always won great prizes. Cortés later told the great cleric Bartolomé de la Casas that his justification for abducting Montezuma was that 'whoever does not enter by the front door is a thief and a robber' – which Las Casas, rightly, took as a prime example of Cortés's thuggishness.

'GOOD COP, BAD COP'

During the weeks of captivity that followed, Cortés played psychological games with his prisoner, alternating threats and cajolery (the modern idiom would be 'good cop, bad cop'). He impressed both Montezuma and his subjects outside the palace with his own fearsomeness by having the 'criminal' Mexica lord at the coast (the one who had struck back when attacked) brought back, condemned for 'treason' and burned alive; obsessed as they were by the magical significance of death by the obsidian knife, the Aztecs would have been profoundly shocked by the death by fire of one of their greatest and best. Yet Cortés alternated these 'proofs of toughness' with courtly forbearance towards Montezuma, even playing board games with him. He dealt severely with anyone rash enough not to show due deference to the royal captive. But he nagged Montezuma mercilessly about Christianity and the evils of human sacrifice. For all that, Montezuma, in an early example of what we should now recognize as Stockholm syndrome, soon grew extremely fond of his abductors and knew them all by name. On one occasion Cortés histrionically removed the chains from Montezuma's ankles and told him he could go free; it is uncertain whether this was play-acting or a genuine offer. But by this time Montezuma feared he had lost credibility with his own people and did not take up the offer.

Cortés meanwhile commissioned his shipwright Martin Lopez to build brigantines to take him, his 300 men and the horses back to the mainland, avoiding the bridges and causeways. Four ships were built and in one of them, such was the superficial cordiality between jailer and prisoner, Cortés actually took Montezuma on a long cruise round the lake. At some point, probably around the turn of the new year, Montezuma actually did promise, this time genuinely, to become a vassal of the king of Spain. This was important to Cortés the shyster lawyer, as it meant any opposition thenceforth could be branded as rebellion and treason.

Life at Tenochtitlán entered a curious limbo. On the surface, the daily routine went on as normal, almost as if nothing untoward had ever happened. Cortés sent out many expeditions to uplift treasure from the Aztec gold mines, whose location Montezuma seems eventually to have revealed. But under the surface there were darker currents. Cacama, the king of neighbouring Texcoco, moved into overt opposition to the Spanish and headed an incipient resistance movement until Cortés, always supremely efficient at espionage, found out about it and had him arrested (it was typical of Cortés that his secret agent on this occasion was Cacama's brother).

The Cacama affair seems to have hardened Cortés's stance against Montezuma. He began to insist that it should be part of a new Mexica constitution that Montezuma and the other caciques must swear an oath of fealty to the king of Spain and become Christians. This was rightly read on all sides as a forced dethronement of the emperor. Cortés was already thinking ahead, trying to outflank Velázquez in Cuba and his other enemies by making his venture 'legal'; if princes of the Church, his most likely critics, ever condemned him for the immorality of his actions, he could then always claim that he had acted to suppress rebellion. Having, to his own satisfaction at least, established an ascendancy over Montezuma, he was starting to turn his mind to the purely Spanish dimension. It is no accident that at this very time his camp was racked by dissension about the Aztec gold. The men were clamouring for an immediate division of the treasure, as rumours were current that Cortés and his captains were secretly salting away stashes before the official division took place. Cortés's treasurer, Mejía, angrily accused Juan Velázquez, Cortés's third in command, of not deducting the royal fifth and Cortés's fifth from his share of the gold, and the two men almost came to blows. Cortés melodramatically imprisoned both of them for a cooling-off period, but Montezuma, who liked Velázquez, offered to compensate him for any losses he sustained in a new division of spoils.

THE LURE OF GOD AND GOLD

Cortés's rank and file certainly had reason for their suspicions about peculation. Financially dishonest along with his other 'sterling' qualities, Cortés collected gold worth 700,000 pesos from the mines and then gave out that the figure was only 160,000. Bernal Diaz famously said that he and the conquistadores had come to the New World to serve God and become rich, and in these limbo days in

Tenochtitlán there was a curious symbiosis between the sacred world and that of Mammon. When not wrangling about the gold uptake, Cortés was engaged in boorish outbursts over Christianity and idolatry. When Cortés revisited the Great Temple and found idols covered in blood from recent human sacrifice, he flew into a rage and attacked an idol with an iron bar, proceeding then to tear off the gold masks that adorned the other idols. He stormed back to Montezuma and peremptorily demanded that the idols be removed and a cross and a temple of the Madonna be put in their place. This was the exact, selfsame demand he had made earlier of Montezuma. The hapless emperor this time offered a compromise whereby both Christian and pagan idols would inhabit the same temple, but the disillusioned Aztecs removed their idols and hid them in a secret location.

By the beginning of March 1520 the Aztecs seemed to have recovered from their initial paralysis. More and more rumours of plots began to surface, and Montezuma himself explicitly warned Cortés to leave before the Spanish were attacked. Cortés asked for carpenters so that he could build an ocean-going fleet in which to return to Europe but then crushed Montezuma's short-lived lifting of the spirits by adding, as if in an afterthought, that it would of course be necessary for the emperor to go with him when he returned to Spain. Montezuma supplied the labourers, who returned to the coast with Martin López, and soon the Atlantic coast rang with the sound of felling trees, sawing wood and hammering nails.

In reality Cortés had not the slightest intention of returning to Spain in the immediate future, for the lure of gold was still strong. But every day there were more signs that the Aztecs had reached the limit of their patience with their white tormentors. The spies whom Cortés had placed close to the person of Montezuma kept hinting, without providing any clear proof, that Montezuma was carrying on secret negotiations with his more warlike chiefs outside the palace, and that a general rising was imminent. Suddenly, there was more dramatic news. The seesaw oscillation between the Aztec and Cuban factors suddenly swung violently in the form of a major expedition sent to Mexico by Velázquez, governor of Cuba.

VELÁZQUEZ'S LONG-PLOTTED REVENGE

Velázquez was finally implementing his long-plotted revenge against his upstart protégé. He sent to Mexico Pánfilo de Narváez, an arrogant Castilian, with a force of 900 men and 18 ships. Many of his comrades were hardened conquistadores known to Cortés, veterans of early campaigns in Hispaniola and Cuba. They had joined Narváez partly out of disappointment with the pickings to be made in the Caribbean islands and partly because the governor obliquely hinted that he would confiscate their land holdings if they did not cooperate with him. The weakness of the new expedition, apart from the feeble personal qualities of Narváez, was the motivation of the men. Loyalty to Velázquez was scarcely a factor, but the lust for riches was, and this was something on which a gold-rich Cortés could play. As soon as Narváez landed at Vera Cruz in early April 1520, Cortés sent an envoy, a friar, to Narváez's camp to tell his lieutenants that their future would be better served by an alliance with Cortés, as he alone knew where all the Aztec gold was located. Initially, though, Narváez did not make any obvious mistakes, and it was when he won the Totonacs over to his side that Cortés realized he would have to deal swiftly and firmly with this new menace if his entire position in Mexico were not to be subverted. Leaving Pedro de Alvarado in command in Tenochtitlán with just 120 men, he set out eastwards for the coast.

At Cholula he was joined by Velázquez de Leon, who had been absent on an exploring mission. With his 260 men, Cortés's force now numbered 350, far less than Narváez's. It was imperative to offset the numerical disadvantage with the element of surprise so, after a rousing speech to his men, the leader outlined his strategy. While keeping up the charade of an exchange of delegations, each offering the other terms, Cortés began a forced march, planning to fall on the enemy unexpectedly. Two days after leaving Cholula he met his envoy, the friar, on the road, and received the most alarming news that momentarily threw him off balance: Narváez, it seemed, had opened a secret channel of communication with Montezuma and the two were colluding to destroy Cortés. Continuing towards the coast via Orizaba and Tanpaniguita, Cortés called a short halt while he distributed gold to his troops and promised them all kinds of benefits once they had defeated Narváez. He sent his friar envoy back to Narváez with a hardline ultimatum, giving up the previous pretence of negotiations: either Narváez must throw in with him and work for the common cause or he would be destroyed. Meanwhile he sent Velázquez de Leon, previously a friend of Narváez's, to his camp to talk to him about coming to terms. Narváez was angry that Velázquez would not put his old friendship for himself before loyalty to Cortés and wanted to arrest him, but was dissuaded. In the meantime Velázquez successfully sowed dissension among Narváez's followers by reading Cortés's terms aloud to the assembled army and talking of the El Dorado that was Tenochtitlán. One chronicler estimated that by the time he left the camp, he had managed to suborn no less than 150 of Narváez's men.

Having sapped morale in Narváez's camp and even bribed his gunners to disable his big guns, Cortés prepared his final stroke. In an eloquent speech he told his troops they now had no choice but to fight or die, since Narváez was treating them like Moors and regarding them as outlaws. The two armies advanced close to each other at Cempoallán, and the rival commanders exhorted their men, offering lavish financial rewards (up to 2000 pesos) to the man who captured the enemy leader. Cortés intended to concentrate his forces on two main objectives: the rapid seizure of the enemy artillery and the apprehension of Narváez. It was the night of 28–29 May. Narváez's host settled down to sleep through a rainy night, expecting to give battle at dawn. But Cortés struck camp soon after midnight and advanced under cover of the rain and darkness. Close to Narváez's positions they were discovered by a sentry who ran back to give the alarm, but in the darkness there was confusion and then it was realized that the big guns were not ready. One of Cortés's commando parties managed to cut the girth straps on the horses, knocking out a possible counter-response by cavalry. It seems that initially Narváez was incredulous about the assault, believing no conquistador army would ever attack another in the midst of savages. He and his bodyguard were quickly isolated by an 80-man company under Cortés's commander Sandoval and in the confused fighting that followed Narváez lost his right eye to a pike thrust and was then captured. Further disarray was caused to Narváez's harquebusiers when they found the rain had made their gunpowder wet. On hearing that their leader had been captured, the never highly motivated bulk of Narváez's men simply laid down their arms.

It was another great Cortés victory, cheaply won, with just two of his own men and 15 of the enemy dead. Some historians have remarked that the battle was won by gold rather than steel, pointing to the successful campaign of bribery and subversion, but the truth is that the newcomers were lacklustre soldiers while Cortés's troops were by now the veterans of half a dozen hard-fought encounters. Having lost their leader, the captains of the ships meekly came to heel, obeyed Cortés's orders, unloaded their copious cargoes of food and then beached the ships in accordance with the victor's commands. Cortés kept Narváez in Vera Cruz as a hostage but simply integrated all the rest of his men and captains in his own force. He now had a much more formidable array to take back to Tenochtitlán.

Cortés had planned a leisurely return to the Aztec capital, but suddenly terrible news came in. The headstrong Pedro de Alvarado, whom he had left behind in command, had escalated the latent conflict with the Aztecs to the point of no return. Before Cortés left, Montezuma had secured his permission to hold the traditional Aztec fiesta of Toxcatl, but the paranoid Alvarado got it into his head that the occasion would be used for a general uprising. To pre-empt this, he went down to the main square with his men and began to massacre all who had assembled there for the festival. He executed Cacama and put other leaders in chains, but the massacre and the brutal treatment of the caciques turned Alvarado's fears into a self-fulfilling prophecy: the Aztecs rose in a general insurrection. Alvarado forced Montezuma to order his people to stop fighting, and for a while they did. But it was apparent to everyone that he had lost his authority and that Alvarado and the Spanish were now under siege in their quarters.

The victorious Cortés had split his forces after the battle with Narváez so had to recall them hurriedly, with orders that all Spanish troops should rendezvous with all speed at Tlaxcala, where a general muster would be held. Tired, short of food, dispirited and despondent, the conquistadores trekked back from the coast through landscapes eerily silent, as if the local tribes were in hiding, waiting to see how the final trial of strength between Spaniard and Aztec would turn out. At Tlaxcala, Cortés mustered an army of about 1000 Spanish troops and 2000 Tlaxcalan allies. He then marched on Tenochtitlán, taking a westerly route round the lake and approaching from the west on the Tacuba causeway. There he learnt that Alvarado and his men were still holding out but in desperate straits. On 24 June the conquistador army re-entered Tenochtitlán, to sullen silence from the Aztecs but great relief from Alvarado and the defenders, who were down to their last few days of food and water. Loyal to his close friends, Cortés did not reprimand Alvarado for his signal folly, but made sure he never again gave him an appointment as his second in command. Rationalizing the fiasco and still brooding over the secret correspondence between Montezuma and Narváez, Cortés at first refused to speak to the Aztec leader, relenting only when he realized that he was now a cipher with no authority and therefore no longer of any significance.

A REMORSELESS WAR OF ATTRITION AND ATROCITY

Montezuma, whether through naiveté or cunning is uncertain, advised Cortés that if he wished to return Tenochtitlán to normal life and reopen the markets, the only way to do this was to release his brother Cuitlahúac. Cortés did so, apparently unaware that Cuitlahuac had always hated him and loathed Montezuma's appeasement policy. From now on the real Aztec emperor was Cuitlahúac. As in the case of Lucknow in the Indian Mutiny in 1857, the relieving force was itself at once surrounded so that a second siege began. An urban guerrilla war commenced: there was continuous low-level warfare, all Spanish foraging and probing parties were attacked, food and water were short and conquistador casualties, especially wounded, mounted daily. Under a constant hail of stones, the Spanish were beginning to realize that their much vaunted artillery was almost useless in the grim business of street-fighting. Veterans of wars in Italy, France and against the Turks conceded that the Aztecs were the toughest opponents they had ever faced. Even worse, having been hoodwinked and bamboozled for six months by Cortés's mind games, the Mexica now had the edge in psychological warfare, grinding down Spanish morale in a remorseless war of attrition and atrocity.

What used to be the conquistadores' trump card was now no more than a busted flush. Despairing of making headway against the guerrilla warfare, Cortés insisted that a reluctant Montezuma go on to the roof of the palace and address his people; it was put to the emperor that his choice was, make a speech or die. It turned out to be no choice at all. His appeals were greeted with a chorus of insults and catcalls, then a hail of stones. A badly wounded Montezuma staggered out of sight, and a frantic Cortés tried to negotiate with the ringleaders among the stone-throwers. He was told unequivocally that the Aztecs would never stop fighting until the Spanish left, though it was by this time far from certain that the white men would be allowed to leave peacefully. On 28 June Cortés probed to see what resistance there would be to an attempted break-out. His worst fears were confirmed when a strong Spanish force was beaten back with heavy losses. Cortés continued to demand that the Aztecs surrender and, for reasons of credibility, launched an attack on the temple of Yopico on the 29th; the temple fell only after the most bloody hand-to-hand combat. Finally, as if to emphasize that the old days were gone for ever, on 30 June, Montezuma died of the wounds he had sustained in the stone-throwing. In retaliation Cortés killed all the high-born Aztecs he still had in captivity.

EXODUS

Cortés had originally been adamant that he would never leave Tenochtitlán but would stay and fight it out. During the terrible week the relieving force spent in the capital, his captains eventually persuaded him that this would be suicide, and pointed to the steady rate of loss: the numbers of Tlaxcalans at their side had dwindled alarmingly, they were short of food and gunpowder, and the enemy daily grew stronger while they grew weaker. Finally persuaded by these arguments, Cortés made elaborate plans for evacuation.

The Spanish would leave the palace of Axayacatl at midnight, and Sandoval and Marina would be in the vanguard, with Cortés and the bulk of the army following closely behind. The Tlaxcalans would form the penultimate column and then would come the rearguard proper commanded by Alvarado and Velázquez de Leon. Since the Aztecs had ripped up sections of the causeway and destroyed the bridges, the conquistadores constructed a portable bridge from the beams of the palace ceiling. The idea was that, as there were three breaches, the bridge would be taken from one to another once the army had passed. As the Aztecs did not like to fight at night, if luck was with them the Spanish should slip away unobserved and be outside the orbit of Tenochtitlán by daybreak. In an evil hour Cortés proclaimed that the men could help themselves to the gold once the king's one-fifth share had been loaded on the pack animals. 'Take what you will of it,' he said. 'Better you should have it than these Mexican dogs.' The more incautious loaded boxes, wallets and even their clothes with bars, coins and ingots.

The straggling army began trooping out of the palace at midnight, the horses' hooves muffled and their exodus masked by the mist and light rain. Cortés's chosen route was over the Tacuba causeway. At first everything went well, and the adventurers even found the first four bridges, in the city proper, intact. But just as they began crossing the lake itself, a Mexica woman spotted them and raised the alarm. Almost immediately the fugitives heard the dreaded sound of the war drum on the summit of the great pyramid. Aztec warriors rushed to their canoes and soon the lake was a seething cauldron of splashing blades as the pirogues shot like arrows towards the Tacuba causeway. The very first contacts showed the Spanish that this was their severest trial yet, for this time the Aztecs were striking to kill, not wound or capture. The engineers managed to lay down the bridge over the first breach, and the vanguard and Cortés's centre force passed over safely; under a fusillade of arrows, they nonetheless managed to reach the mainland near Tacuba, though they had to swim across the last two channels as the causeway was broken. Cortés on horseback and five mounted companions then turned back to help their stricken comrades in the rearguard. By now the army, marching probably 15 abreast at most, was hopelessly strung out along the causeway, with the rearguard trapped between the first and second breaches. The portable bridge had become stuck fast in the side of the dyke, the timbers pressed down into stone and earth so that the different molecules became almost indistinguishable. Meanwhile the Aztecs, climbing up the sides of the causeway, were massing in strength at the second breach, cutting off the Spanish rearguard.

The Night of Sorrow or La Noche Triste

The would-be rescuers found scenes of unutterable chaos and made little headway. Many soldiers had already been killed or drowned, and the entire causeway was being assailed by myriad canoes. Some of the wounded gave up the fight and dropped into the water, and others panicked, became paralysed by fright or otherwise disabled and committed suicide by drowning rather than be taken alive and hauled away for sacrifice. One eyewitness wrote later that there was not a man who would lift a hand to help his companion nor even his own father or brother. In a nightmare scene of midnight horror it was every man for himself. The cannon, many of the horses and most of the gold were lost very early in the battle. The gold was a true curse, since men were slowed down by their eagerness to escape with it, and it was noticeable that this particularly applied to the men who had come with Narváez. Blood was everywhere, men screaming in death throes, horses whinnying in agony, the Mexica warriors ululating blood-curdlingly and the throb of tocsin and drum sounded a macabre counterpoint. Panic was a worse enemy than the Aztecs, for no attempt was made to organize a fighting resistance or any kind of last stand. Terrified men simply trampled each other and thrust their comrades aside to get nearer to the breach or jump into the water.

The main disaster was at the Toltec canal, the second cut in the causeway after leaving the city. So many men fell here that for a time the breach was filled with a mass of dead bodies, so that those who came last were able to step over them as if on a solid foundation. Cortés himself came close to capture when he fell into the water and would have been seized and carried off for sacrifice but for the valiant intervention of two of his bodyguards. He actually did manage to find a fordable path through the water and exhorted his men to follow by that route, but in the hellish din no one heard him. Soon he and the handful who had hacked their way through the inferno from the rearguard plunged back into the water separating them from the mainland; more died here, and the survivors came through mainly by holding on to the manes and tails of the horses.

Alvarado, in command of the rearguard, managed to save himself in controversial circumstances – once again Cortés did not enquire too closely into the conduct of his favourite. But his co-commander Velázquez de Leon was never seen again: either he perished in the fighting or was sacrificed later. Very few of the rearguard made it to safety, and Spanish *esprit de corps* was hardly reinforced by the disaster. According to one report, 270 Spaniards stationed in the outer reaches of the palace were not even told of the decision to evacuate and were then surrounded, captured and dragged away to sacrificial death by the obsidian knife. By any standards the events of what came to be known as La Noche Triste (The Night of Tears) were a catastrophe and by far the greatest set-back to Spanish arms since the discovery of the New World. More than 600 Spaniards perished that night, and nearly all the Tlaxcalans. Cuitlahúac, hailed as deliverer and redeemer, was credited by the Aztecs for being the mastermind behind the bloody retribution against the tormentors, though the battle was in truth a spontaneous killing spree, when the Mexica for once had the Spanish at their mercy. He was at once crowned as the new emperor and promised his people he would rid the land of this foreign plague once and for all.

THE VICTORY AGAINST CUITLAHÚAC

Returning disconsolately to the mainland, Cortés displayed stoical sangfroid (or ruthless indifference). Showing no signs of panic, he told his men that he still intended to conquer Tenochtitlán and that they had simply lost one engagement in a long war. He asked personally after only one person – Martin Lopez, the shipbuilder – and once he heard he was safe, remarked laconically: 'Well, let us go, for in that case we lack nothing.' His grim self-satisfaction was no doubt a product of the news that most of those killed had been Narváez's men. But the Spanish were far from safe yet. As Cortés retreated north along the lake from Tacuba, he was under constant attack from the pursuing Aztecs. He kept his men marching all day, then made a brief stop at nightfall for sleep before setting off again at midnight. Swinging east at the north of the lake, with their ultimate destination Tlaxcala, the conquistadores must momentarily have thought they had escaped lightly, for the attacks of the pursuers now seemed sporadic and uncoordinated. But Cuitlahúac was merely lulling his prey before launching his main attack. Maybe a week after The Night of Tears, at Otumba, the Aztec host came upon the Spanish army, now reduced to about 340 men, nearly all wounded; about 400 had reached the mainland at Tacuba, but 60 more had died of wounds as a result of the gruelling marching.

Otumba may well be Cortés's greatest military exploit. The battle raged all morning, with the confident Aztecs ranged against the wounded and demoralized conquistadores. But the Mexica had reverted to their old mistake of trying to take prisoners for sacrifice later instead of aiming to kill. Reduced to 27 horses – three had died on the trek round the lake – Cortés mounted his most able warriors and charged straight at the caparisoned Aztec leaders clustered around the standard. Quickly overcoming the resistance and seizing the banner, the Spaniards sowed confusion in the enemy ranks, as the loss of the rallying point both depressed and confused the Mexica warriors. Cuitlahúac demonstrated that he was not a mighty warrior after all. Not only had he ordered his men to take prisoners, instead of unleashing the natural homicidal fury in evidence during La Noche Triste, but he had also made the elementary mistake of attacking mounted men in open country. The Aztecs were deadly in street-fighting or urban guerrilla warfare but they had no answer to the sudden cavalry charge at which Cortés excelled. Nor could they deal with the Spaniards' secret weapon, most notably in evidence at this battle, the ferocious, slavering mastiffs trained to go for the throat. If the victory against Narváez owed a lot to gold, the victory against Cuitlahúac owed much to animals – horses and dogs.

The conquistadores reached Tlaxcalan territory on 9 July and, to their great relief, found their old allies friendly. Knowing how volatile the Indian tribes could be they had feared that the Tlaxcalans might not hold firm, and indeed their fears were well grounded, for the decision to support them had been touch and go. Cuitlahúac had sent envoys to Tlaxcala to try to persuade them to abandon the defeated Spanish, and the strong anti-Cortés faction in Tlaxcala had nearly won the day, only to be narrowly outvoted in the end by the Cortés loyalists. Hatred of the Aztecs had overruled their true self-interest. Even so, the Tlaxcalans drove a hard bargain. Cortés had to agree to hand over Cholula, to man a new fortress in Tlaxcalan territory to protect them against an Aztec revival, to pledge a guarantee of perpetual freedom from tribute from whoever ruled in Tenochtitlán and to divide all booty with them.

The machiavellian Cortés agreed, knowing that if he defeated the Aztecs, he could easily repudiate the agreement and there was nothing Tlaxcala could do about it. He had more problems with his own men, for many of them did not share his confidence in an eventually successful outcome and clamoured to return to Cuba. Cortés once more proved himself a genius with persuasive words. He appealed to the Castilian ethos of honour to get them to stay. His speech to the disaffected contained some notable points: he claimed he could not break his word to the Spanish king; if they retreated, the Tlaxcalans would throw in their lot with the Aztecs; he argued that all conquerors suffer one defeat (not strictly true) and that no great captain gave up after losing just one encounter.

IN PURSUIT OF A GOLDEN DREAM: TOWARDS TENOCHTITLÁN

In the meantime the Aztecs had not capitalized on their great victory on The Night of Tears. Thinking the Spanish had left for good, they became overconfident, repaired their city and returned to their traditional feasts. Almost immediately they ran into set-backs. Smallpox struck them in September 1520 and though the impact of this should not be exaggerated (as it has been by some historians), psychologically, it hardly seemed that their gods were favouring them, especially when Cuitlahúac himself died of the disease shortly after being chosen as emperor. Cuauhtémoc, a more able man, was elected in his stead. To the dismay of the Mexica, they learnt that the Spanish (with their immunity acquired in Europe) did not succumb to smallpox as they did. Nor was the Aztec search for new allies very successful, and Cortés compounded the problem by striking at the people of Tepeaca, one of the most steadfast tribes in the Mexica coalition. Cortés's action was shrewd, as it hit a number of targets in one: it demoralized the Aztecs by alerting them to the speed of Spanish recovery; it impressed the Tlaxcalans; it cut the ground from under the faction that wanted to return to Cuba, and it concentrated the minds of the many fence-sitters.

More ships arrived at Vera Cruz to consolidate the Narváez expedition, so Cortés simply incorporated their complements in his own forces. By the time he set out for Tepeaca, 65 kilometres (40 miles) southwest of Tlaxcala, he had 500 infantrymen, 17 horses and 6 crossbowmen. Early in August he won a fierce battle and then spread a reign of terror throughout Tepeaca province. By this time he was increasing his atrocity count: for the first time he sold women and children into slavery and ostentatiously turned a blind eye to human sacrifice by the Tlaxcalans, which he had previously deplored. His most brutal campaign yet secured rich dividends. He inspired fear everywhere, won new allies, secured a solid base of operations and cut off the Aztecs from both the sea and important supplies in the interior. Some have speculated that after La Noche Triste the iron entered Cortés's soul, and he became a genuine monster in his treatment of the Indians, but it may simply be that he no longer needed to dissemble, as in the earlier, more tentative days in Mexico. And every day his strength increased. Six more ships arrived at Vera Cruz, some sent by governor Velázquez in ignorance of what had happened to Narváez; at least 200 more men were added to the army. The hardcore of early Cortés adventurers was now just one of three distinct groups of conquistadores, the others being the surviving Narváez men and now the newcomers.

At last Cortés was ready to fulfil his golden dream: the taking of Tenochtitlán. Advancing with a formidable host, he arrived at the lakeside city of Texcoco on New Year's Eve 1520. He had 550 Spanish infantry, 80 crossbowmen and harquebusiers, 40 horses, 8 field guns and 10,000 Tlaxcalan auxiliaries. Texcoco was a virtual ghost town, since most of the inhabitants had fled, but Cortés rested there three days before marching round the south of the lake to Ixtapalapa. The Aztecs showed their increasing resourcefulness, devising a scheme to drown the Spanish army by diverting flood water into the town but, as luck would have it, the conquistadores had camped outside the town so were unscathed by the mini-tsunami. Until the end of January Cortés was busy, hauling 12-metre (40-foot) brigantines over the mountains to the lakeshore, ready for an amphibious assault. Sea power would be all important in the coming struggle, and Cortés blessed the good fortune that had made him preserve the ironwork and rigging of his beached armada at Vera Cruz.

Cortés set out west around the lake as far as Tacuba where, in a six-day stay, he fought many bloody skirmishes; there were more such encounters, too, when he retraced his steps to Texcoco. There he had to quell yet another conspiracy against him (by an agent of the Cuban governor named Antonio de Villafana) and execute the ringleaders. In three months of sustained warfare, he sent out expeditions to outlying areas like Cuernavaca. Meanwhile, all around the lake there were fierce battles as Cuauhtémoc sent waves of warriors over the lake in a counteroffensive. At one battle at Xochimilco Cortés, a cat of many lives, would certainly have been killed if the Aztecs had been fighting to slay rather than capture. By the end of the three months Tenochtitlán was effectively surrounded, short of food and supplies, receiving little tribute and forced to make war in the planting season.

HISTORY'S LONGEST NON-STOP BATTLE

As the smallpox was making inroads on the Mexica population, all this time the Aztecs were growing weaker and the Spanish stronger. Cortés was slowly learning new techniques, such as how to use

216

crossbowmen against attacks by water and how to deploy horsemen on the causeways. By the end of April, with constant reinforcement, he had 700 infantry, 120 harquebusiers and crossbowmen, 90 horsemen, 3 large guns and another 15 smaller ones mounted on the brigantines. On 28 April he launched 12 brigantines on the lake, a sign that he had now gained mastery on water as well as land. His strategy was very clear: starve the city, destroy all war canoes, occupy the causeways, and do not waste men needlessly; as was his usual custom, Cortés always tried to get the Tlaxcalans to take the brunt of the early fighting, using the conquistadores as shock troops. Tactically, he divided his forces into four groups, three of which would fight on land and the fourth as marines on the brigantines. In the initial stages of the campaign, the three land armies were to seize and hold the Coyoacán, Tacubs and Ixtapalapa causeways while the brigantines cleared the lake. On 1 June Cortés ordered his brigantines to advance across the lake; meeting the Mexica war canoes half-way across, they won a shattering victory, with the bronze guns, the harquebuses and crossbows all contributing their share to the destruction of the enemy. Cortés now had mastery of the lake. The Coyoacan contingent under Cristóbal de Olid followed this up with a major land success.

Cuauhtémoc ordered a grand mobilization of the Aztec people – the first time their society had been organized for total war rather than the campaigns of a warrior class. Now women too took up arms, finally confirming the legends about 'amazons' so pervasive in the Spanish conquest and responsible for the name of South America's largest river. The Mexica were already hard pressed, since early in June the Spanish disabled the aqueduct into the city; with the main water supply cut off, the defenders of Tencochtitlán had to rely on wells. But weeks of fighting at first produced no real progress, partly because of the ferocity of Aztec resistance and partly because Cortés, jittery about Spanish casualties, let the Tlaxcalans do most

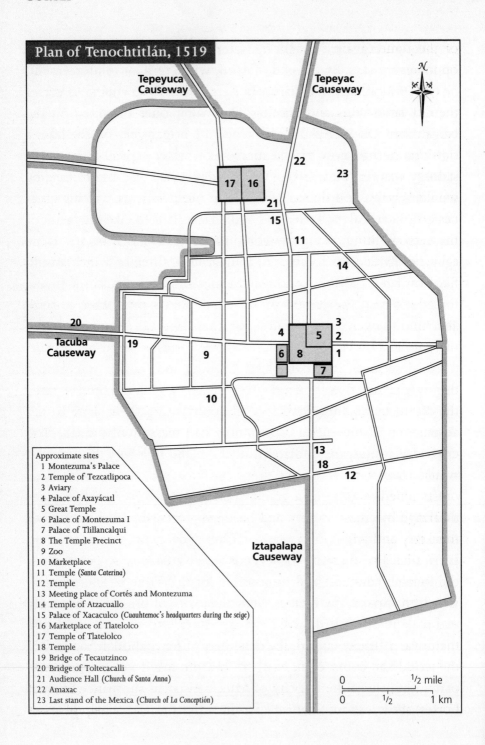

Plan of Tenochtitlán, 1519

Tepeyuca Causeway

Tepeyac Causeway

N

22

23

17 16

21

15

11

14

20

Tacuba Causeway

19

4

3

5

2

6

8

1

9

7

10

13

18

12

Iztapalapa Causeway

Approximate sites
 1 Montezuma's Palace
 2 Temple of Tezcatlipoca
 3 Aviary
 4 Palace of Axayácatl
 5 Great Temple
 6 Palace of Montezuma I
 7 Palace of Tlillancalqui
 8 The Temple Precinct
 9 Zoo
10 Marketplace
11 Temple (*Santa Caterina*)
12 Temple
13 Meeting place of Cortés and Montezuma
14 Temple of Atzacuallo
15 Palace of Xacaculco (*Cuauhtemoc's headquarters during the seige*)
16 Marketplace of Tlatelolco
17 Temple of Tlatelolco
18 Temple
19 Bridge of Tecautzinco
20 Bridge of Toltecacalli
21 Audience Hall (*Church of Santa Anna*)
22 Amaxac
23 Last stand of the Mexica (*Church of La Conceptión*)

0   1/2 mile
0   1/2   1 km

of the fighting and, man for man, they were no match for their opponents.

On 10 June, Cortés committed the Spanish to attack the centre of the city across the causeway, but they were driven back by a ferocious counterassault; already the non-stop fighting was reducing the outskirts of the city to rubble. Cortés ultimately realized, with some sadness, that, in the words of a famous phrase from later history, he would have to destroy the city in order to save it. The sadness was not so much for the architectural splendour of Tencochtitlan as for his own credibility; he had wanted to deliver a beautiful city to the agents of Charles V of Spain but now faced the reality that he would hand over merely a smoking ruin. A similar assault on 15 June reinforced the message that the Aztecs would fight to the death, that all Tenochtitlán would be consumed in the holocaust and that none of the gold lost on The Night of Tears would ever be recovered.

No description of the ferocious, bloody, house-to-house combat during the struggle for Tenochtitlán could ever adequately convey the diabolical quality of a hell on earth. Typically, the Spanish would advance up a street until they came to a large ditch or canal, once crossed Venice-style by a bridge long since destroyed. Here the Aztecs would make a stand behind stone barriers that bristled with their outstretched spears. Inevitably, the conquistadores would then order a barrage by crossbowmen and harquebusiers, but when the roll of musketry and whiz of arrows made no impression against the stone barricades, the big guns would be called up. Only when a ferocious cannonade blew holes in the barricades would the Aztecs retreat, at which the Spanish would run forward, leap into the shallow water and scale the defences on the other side. Cortés himself wrote that the noise of battle was so deafening that it sounded as if the end of the world had come. The cacophony had myriad elements: the noise of gunfire, the blood-freezing Mexican war cries, the neighing of horses, the lamentations of the wounded, the cries for help of those being dragged away for sacrifice, the roar of gutted buildings collapsing, the shrieks of women entombed by falling masonry and the screams of those whose throats were choked with dust from the rubble and crying out for water. Some historians have claimed that the struggle for Tenochtitlán was the longest non-stop battle in all human history and certainly only the titanic contest for Stalingrad in late 1942 can match it for sustained horror. Everywhere the conquistadores went they encountered stakes, barricades, ramparts, hidden pits and other earthworks. It seemed as if rain and hailstones pelted down continually, such was the barrage of stones, darts and hissing arrows.

The Aztecs were proving to have a superlative talent for street-fighting. Any initial advantage in warfare provided by Toledo steel had been negated by the piles of Spanish weapons captured in more than a year of fighting. Cuauhtémoc's men now fought with lances tipped with captured Castilian blades as well as the swords of the conquistadores. Concealed on rooftops, their rock-slinging snipers were brilliantly effective, and it became clear that to flush them out the Spanish would have to destroy every single house and leave not a stone on a stone. Cuauhtémoc kept hoping that the Spanish would make the cardinal error of establishing a base in the centre of Tenochtitlán, when he might hope to surround and massacre them, but Cortés was too shrewd for that; he pointed out to his captains that, with a permanent base in the centre, they would have to fight all night and every night. Instead, he tried to grind down the enemy by a campaign of attrition, constantly entering and leaving the city on fresh forays, never making a fixed headquarters there. Yet every time the Spanish came close to the centre of Tenochtitlán, they had to answer with their lives. They had to destroy every single building in a street before venturing down it, for if they left the houses intact, the Aztecs would reoccupy them when the Spanish withdrew at night, ready to rain down a fusillade of stones and arrows next morning, so that the selfsame streetfight had to be re-enacted next day.

The courage and tenacity of the Aztecs seemed beyond belief. Every time Cortés thought he had finally broken the back of the enemy, there would be a fresh set-back. Overconfidence always brought its own nemesis. On one occasion, after a fierce combat in the main square, the conquistadores thought they had the enemy on the run and pursued them into the temple enclosures, only to be driven out again by superior numbers, the Aztecs having concealed their reserves there. Every inch of ground had to be fought for. The Spanish carried pontoon bridges to lay across the broken causeways and employed the Tlaxcalans to fill in with adobe and timber the breaches the Mexica had made, but every morning there would be fresh holes and breaches. Aztec bands would appear in the Spanish rear, and Cortés would suddenly have to deal with frenzied attacks in three different sectors simultaneously. Steadily Spanish casualties mounted. Needless to say, Alvarado's judgement was again at fault, and on 23 June a bad mistake led to five of his comrades being captured and dragged away for sacrifice. Even worse, a few days later a brigantine grounded in one of the back canals in what was officially supposed to be conquered territory, and Aztec commandos dragged away another 15 to appease their savage gods on the stone altars.

Nonetheless, by the end of June final victory for the Spanish seemed close, with all non-Mexica tribes having come over to Cortés and the Aztecs withdrawn to their final redoubt at Tlateloco, a northern suburb. On 30 June Cortés gave the signal for an attack by Spanish and Tlaxcalan troops, to converge from all points. It says something for the determination with which the Aztecs were defending their city that Cortés had to overrule his own captains in council to order this attack; a majority of them had voted against the idea, fearing that Cuauhtémoc, who had proved himself a highly talented tactician, would somehow manage to trap them there. On this occasion the despised majority proved better prophets than the leader, so Cortés simply rewrote the episode in his memoirs, claiming that he had opposed the attack but his lieutenants had insisted. The truth is that Cortés used a plan of which he was very proud but that went disastrously wrong. Entering a now deserted Tenochtitlán in his usual way on the morning of 30 June (almost a year to the day after The Night of Tears, Cortés's troops marched into the main square, scene of so many battles already. Then he split his main assault force into two. Julian de Alderete, commanding one column of 70 infantrymen and 8 horse, was to strike northwest at Tlateloco while Cortés proceeded due north along a narrower road to face the enemy in a pincer movement.

The attack began to go wrong almost immediately. Cortés had not factored into his calculations the narrow serpentine streets of Tlatecolco, used as he was to Tenochtitlán's broad avenues. Having made unwontedly slow progress, he then ran into heavy opposition, so decided to abandon his original plan, backtrack and follow Alderete into Tlatecolco. Unfortunately for him, the Aztecs, in a brilliant commando raid, had cut a large breach in the causeway as soon as Alderete's column passed; the gap had only just been filled by the Tlaxcalans but now gaped there again and there was a 11-metre (12-yard) trench full of water 2.5 metres (8 feet) deep between the two columns. Forced back by unexpectedly stiff opposition, Alderete's men suddenly found themselves being tipped into the water, as the fresh breach created widespread chaos. Unable to use their horses or guns in the crowded streets, the Spanish were trapped and, quickly grasping their advantage, the Aztecs jumped into their canoes and paddled round to the rear of the conquistadores. As on La Noche Triste, Cortés tried to help the beleaguered contingent and, on this occasion too, came within a whisker of being taken by the enemy; only the usual Aztec insistence on capturing rather than killing saved him, and this time it additionally required the valour of a Castilian swordsman named Cristóbal de Olea, who sacrificed his life for the commander.

THE AZTECS' LAST HURRAH

The Spanish suffered outright defeat, with 20 Castilians killed in the furious mêlée around the breach and, according to some versions, 2000 of their Indian allies. The most macabre fate awaited the 53 conquistadores captured in the fray. The Spanish leaders had to watch from a distance as their comrades were taken up the steps of the great temple at Tlatecolco, to the sonorous accompaniment of drums, trumpets, horns and conches to be sacrificed. Forced to dance before the idols, the captives were thrown on the altar and had their chests sawn open with obsidian knives; their still beating hearts were then held bloodily aloft and the corpses given to the butchers to be cut up. This was the worst defeat the Spanish had sustained since The Night of Tears, and it had a dramatic effect. The Indian allies immediately began to desert the conquistadores, and for four days a gloomy Cortés stayed in his camp, closely guarded by the brigantines. Great though he was as a warrior, Cuauhtémoc proved to be no diplomat, and in this crucial period he lost his opportunity to raise Mexico against the white invaders. Partly to strike at a soft target, partly to restore credibility, Cortés led a diversionary attack to Cuernavaca against the Otomi, who broke with him immediately after the disaster. The success of this campaign once more restored the initiative to the Spanish.

The great victory on 30 June turned out to be the Aztecs' last hurrah. By now even their seemingly limitless numbers had been thinned by the horrendous death tolls of a solid month of battle, they were exhausted physically, mentally and morally and seriously short of both food and water, since the Spanish brigantines still controlled the lake. When the struggle was renewed in mid-July – after the breathing space provided by the Cuernavaca campaign – the Aztecs immediately seemed less effective; there were no more breaches in causeways or roads. With the rainy season beginning, they had come to the limit of their endurance, and the Spanish delivered the *coup de grâce* by seizing the last fresh-water well. With reinforcements and fresh ammunition for Cortés arriving daily from Vera Cruz, even Cuauhtémoc conceded that this was the end of the road. From 15 July the conquistadores enjoyed a continuous run of success, including the symbolic capture of the temple of Tlatelolco. Yet still the fighting continued. Cortés and Cuauhtémoc both ended up in the position of 'doves' outflanked by hawks and hardliners. Increasingly desperate, brutal and bloody fighting continued until the Tlaxcalans finally slaked their blood lust with a great massacre of surviving Mexica on 12 August. The final battle was fought the next day.

Cuauhtémoc was captured and brought in chains before Cortés, who kept him prisoner for a time. The defeat of the Aztecs was the signal for genocide by the crazed Indian allies, who killed everything in sight in their pent-up hatred of the Mexica. It is thought that 100,000 Aztecs were slain in the conquest of Mexico. The Spanish lost at least 100 dead in the nearly three-month siege of Tenochtitlán, over 1000 in all the fighting against the Aztecs and about 1800 in all in the years 1519–21. Atrocities were overtaken by renewed gold fever, and Cuauhtémoc was nearly tortured to death by his guards to get him to reveal fresh sources of the yellow metal. Because his men were greatly dissatisfied with their paltry pay-off of 50–60 gold pesos each, Cortés was forced to appease them by granting them lands together with Indian labour, which meant that he in effect introduced the detestable Encomienda system to mainland America and condemned the Amerindian population to serfdom. Of Cortés's financial dishonesty there can be no doubt, for the treasure he sent back to Spain to win over Charles V against his many critics and detractors shows clearly that he had lied to his men and possessed far more gold than he had admitted. This was always the Cortés method – to note everyone of importance in Spain and bribe them with Mexican gold. It is a final irony that the gold he sent back to the Spanish courts to win friends and influence people never got there, being intercepted near the Spanish coast by the notorious French corsair Jean Fleury.

A GREAT BUT CONTROVERSIAL CONQUEROR

Cortés was a great conqueror but his methods were always controversial, and he made powerful enemies. Juan Rodríguez de Fonseca, Bishop of Burgos, was at the time of the conquest Charles V's chief adviser on 'the Indies', and he saw right through Cortés and his duplicity. He sent an inspector-general named Cristóbal de Tapia to Mexico in 1521 with plenipotentiary powers, but Cortés, backed by the might of his army, simply ignored him. He relied on court factionalism at Madrid, the plethora of agents he employed in Spain with Aztec gold to lobby for him, the shower of paperwork with which he deluged the court and, most of all, the sizeable quantities of treasure he continued to ship to the Old World to see him through. His instincts were correct. Although Cortés's treatment of Tapia was treason in anybody's language, the Spanish monarch, bored by the New World anyway, allowed himself to be gulled and placated by gold and Cortés's constant protestations of loyal service; it may have been put to Charles that it was more trouble to get rid of Cortés than to humour him, for an express repudiation by the Spanish court would almost certainly have led to a unilateral declaration of Mexican independence by the supposedly loyal subject. Fonseca lost favour in Madrid, Charles V appointed a tribunal to decide the issues between Cortés and governor Velázquez of Cuba and the tribunal, packed with pro-Cortés ciphers, issued a decree (dated 15 October 1522) naming Cortés governor and captain general of New Spain.

Only in 1526 did wiser counsels prevail and, with more and more disgruntled veterans of the conquest accusing Cortés of ingratitude and embezzlement, he was suspended from his position as governor. He returned to Spain in 1528 to seek redress, was given a magnificent welcome by Charles V, but was not reappointed as governor. When he returned to Mexico in 1530 the properly appointed Spanish colonial officials kept him at arm's length, and he spent an unhappy decade defending himself against charges of profiteering and murder. In 1540 he returned to Spain and lived in retirement in Seville until his death in 1547.

Even though morality is not the first quality one looks for in great warriors, Cortés was singularly devoid of it. His defenders instance his apparently devout Catholicism, but he never allowed Christian doctrine to stand in the way of his own self-interest, and his support of religion against the paganism and idolatry of the Amerindians can always be seen to have a clear political motivation. It is a moral certainty that he was guilty of murder outside the wartime context in which convenient executions could be at least argued for, if not justified. A frenzied womanizer, once reluctantly reunited with his wife Catalina Suarez, he ensured her sudden demise. We know that she died after a row over the harem of informal mistresses Cortés kept, and the likelihood is that the great conqueror struck her a mortal blow while in a blinding rage. Cortés's defenders lamely talk of a 'weak heart' but no such extenuating circumstances can be found in the case of the other mysterious deaths of those who crossed him. Julian de Alderete fell out with him and decided to return to Spain to lay before the king the full facts behind the conquest of Mexico; he died suddenly of poisoning. Francisco de Garay, governor of

Jamaica, was perceived to be Cortés's rival in the New World and expired suddenly after dining at Cortés's table. In 1526 the same thing happened to Ponce de Leon and Marcos de Aguilar, both men who had been sent out from Spain to investigate him. The list of mysterious deaths around Cortés thus comprises Catalina Suarez, Francisco de Garay, Julian de Alderete, Ponce de Leon and Marcos de Aguilar. It would be interesting to know the actuarial statistics regarding the likelihood of all these important contacts of Cortés dying a natural death before 1526. Yet the caudillo of New Spain got away with it, just as he had diced with fortune over and over again in his life-and-death struggle with the Aztecs. Cortés lived by the motto 'fortune favours the brave', but the real lesson to be drawn from his life is a more cynical one: nothing succeeds like success.

CHAPTER 5

Tokugawa Ieyasu
Japan's legendary Shogun

TOKUGAWA IEYASU WAS the founder of the Tokugawa Shogunate, a
dictatorship that ruled Japan until the Meiji Restoration of 1868.
The first 60 years of his life were the most turbulent in all Japanese
history, an era of almost constant civil war. Using a broad-brush
approach we can say that those 60 years saw the transformation of
Japan from a village-based society to a centralized autocracy. The
greatest lord, or shogun, commanded the allegiance of lesser, terri-
torially based but still very powerful, lords who in turn dominated
their territories with castles and forced all samurai (warriors) to
serve as their personal retainers. The village used to be a microcosm
of Japanese society, with all classes and ranks represented there, but
the rise of the territorial lords or damyos made them into mere rice-
growing appanages. Two kinds of struggle went on simultaneously:
between the territorial lords and other such landed magnates striv-
ing to be the supremo or shogun, and between the territorial lords
and the other threats to their power within their lands – landless
samurai, peasant revolts, fanatical religious sects. The second half of
the sixteenth century saw nearly all samurai owing formal allegiance
to their territorial lords, and most territorial lords owing allegiance

to whoever it was who had greatest power at the centre at any given moment. Japanese history proceeded at a different pace from that of the West, which is why historians using western models are often baffled. On the one hand, Japan experienced a concentration of centralized autocratic power like that of the Tudors in England. On the other, only now for the first time was there anything like western feudalism in Japan, replacing the previous local, patrimonial and kinship-based structures.

A PAWN IN THE STRUGGLE FOR MASTERY

None of this needs matter to us, concerned as we are with warriors, except for one thing. If the evolution of Japanese society in the years 1550–1600 is confusing to us today, how much greater must the uncertainty, unpredictability and general chaos have appeared to those actually caught up in the maelstrom of marching armies, changing political alliances, treachery, treason and brother set against brother in civil strife? Ieyasu was one such: he must never have known a calm day from the dawning of consciousness to adulthood. He was born on 31 January 1543 in the province of Mikawa in eastern Japan and was originally named Matsudaira Takechiyo (the first name in Japanese always denoted the clan origin). His father was Matsudaira Hirotada, who died at 23 when his son was 6, and his mother O-Dai-No-Kata, the daughter of a neighbouring lord; his parents were actually stepbrother and stepsister, aged 17 and 15 respectively when their son was born (to avoid confusion about Japanese patronymics we shall call him Ieyasu throughout). Two years later Hirotada sent his wife back to her family and never saw her again. Both parents remarried and had further children, so that Ieyasu ended up with eleven half-brothers and -sisters.

One of the early struggles for mastery in eastern Japan was

between the Oda and Imagawa clans, and here the chaos principle first asserted itself, for one section of the Matsudaira family opted to be vassals of the Imagawa and the other plumped for the Oda. Ieyasu's early years saw constant warfare within his own clan between the Oda and Imagawa clans as part of the wider clash; his father had favoured the Imagawas but most of the Matsudairas were Oda partisans. In 1548 when the Oda invaded Mikawa province, Ieyasu's father, Hirotada, turned to the head of the Imagawa clan for help. He agreed, on condition the five-year-old Ieyasu was sent to his castle at Bumpu as a hostage; all Japanese lords routinely feared treachery and hostage-taking was a normal part of warfare. But when the Oda clan heard of this, they ambushed Ieyasu's caravan and abducted the boy. They then informed his father that the lad would be executed unless Hirotada abandoned the Imagawas; Hirotada refused, saying that the death of his son would show the Oda how serious he was. Cleverly, the Oda leaders did not kill Ieyasu but kept him as a pawn for three years at their headquarters in Nagoya. Hirotada died mysteriously at 23 and, when the Oda were hit by an epidemic (type unknown), the Imagawa took their chance, as they thought, to finish them off. Laying siege to the principal Oda castle, they promised to lift the blockade if Ieyasu were given up; the wisdom of the original policy of mercy was thus confirmed. Now aged nine, Ieyasu was surrendered and taken to Sumpu castle where he lived until the age of 15, once again as a hostage.

While still retaining the patronymic of the Matsudaira clan, Ieyasu changed his name twice more in the late 1550s, the second time when he married, aged 16. Deeming him now to be of warrior age, the Imagawa allowed him to return to his native Mikawa on condition that he raised a force to fight the Oda. He soon gained his spurs, winning a minor battle at Terabe and then attracting attention by a daring night attack that allowed him to deliver supplies to a border fort. But he was fighting on the losing side, for in 1560 the leadership of the Oda passed to the brilliant Nobunaga, destined to be one of the 'big three' in Japanese history in the late sixteenth century. Ieyasu was always lucky, and fortune was with him now. The Imagawas invaded Oda lands with an army of 20,000 but were annihilated by Nobunaga at the battle of Okehazama. Ieyasu, detached to capture a border fort, stayed on the frontier to defend it so took no real part in the hostilities. The Imagawa leader who had bound him with oaths had died at Okehazama and Ieyasu, reading the runes correctly, declared himself absolved from all previous allegiances and made submission to Nobunaga. It was typical of the double-dealing that characterized all political and military activity in Japan at this time that he did so secretly, because his wife and infant son were being held hostage by the Imagawa at Sumpu. Only in 1561, when Ieyasu captured the fortress of Kaminojo, was he able to buy back his wife and son by exchanging them for the wife and daughter of Kaminojo's castellan.

A CLEAR POLITICAL STRATEGIST

Now free from all ties, and at the age (18) when an ambitious political leader needed to start making his mark, Ieyasu acted as a model vassal to Nobunaga and spent many years reorganizing and reforming his own Matsudaira clan, pacifying the turbulent province of Mikawa and apportioning lands to his key retainers and vassals. His only armed excursion at this stage in his career was to put down a rebellion of warlike monks in Mikawa — typical of the many uprisings with which a territorial lord had to contend; in the course of the fighting he was nearly killed when a bullet hit him but did not penetrate his armour. Firearms, introduced by the Portuguese in 1542, were the new factor in Japanese warfare, one that Ieyasu followed attentively. Growing in confidence, Ieyasu in 1567 finally adopted the name by which history knows him, calling himself Tokugawa Ieyasu (having received permission from the emperor) and claiming descent from the prestigious Minamoto clan, entirely without proof or foundation; he left the name Matsudaira for cadet families in his clan or to reward cooperative smaller lords.

Japanese lords had to prove their noble ancestry and often kept two different genealogical charts, just as modern fraudsters keep two sets of account books. Ieyasu had two such charts: one linked him to the imperial bureaucracy by claiming descent from the Fujiwara family and the other to the great military family of the Minamoto. He then set about a programme of expansion, always taking care that he had Nobunaga's approval for his endeavours and was thus protected on his western flank. In 1568 he and Shingen, head of the Takeda clan in Kai province, formed an alliance to conquer the Imagawa lands; two years later he had added Totomi province to his domain while Shingen captured Suruga province, including Sumpu castle. Fearing that Shingen had now become too powerful, Ieyasu then abandoned him and made an alliance with the Kesugi clan – the hereditary enemies of the Takeda.

By now Ieyasu's political strategy was clear. First, follow the power and do not alienate the most formidable in the land, which for the moment meant Nobunaga, who was already ascending to supreme mastery in Japan. Nobunaga's career showed the importance of strategically located power bases. His was Owari and the Kinai, which had several advantages: central authority was always weak there but, paradoxically, it was the political heartland of Japan, with the highest levels of productivity, and Owari had fertile fields abutting the great river basins such as the Kiso river delta, where advanced flood-prevention techniques enabled sophisticated self-governing villages to thrive. Ieyasu was a quick learner and the lessons he derived from Nobunaga's rise were to consolidate one's power base, ensure that you have a rice supply to feed a large army and the infrastructure to resolve logistical problems, and always look over your shoulder at rivals. It was crucially important never to challenge the dominant power or 'hegemon' prematurely, which was why Ieyasu made a point of leading 5000 men into battle at Anegawa in 1570 when Nobunaga defeated the Asai and Asakura clans. Ieyasu's deference towards Nobunaga paid off when his previous ally Shingen, now allied with the Tojo clan, attacked Ieyasu's lands in Totomi in 1571.

But this was not Ieyasu's finest hour. As if to illustrate the perennial motif of treachery and double-dealing in Japanese history of the time, the 3000 men Nobunaga sent to Ieyasu's aid deserted, allowing Shingen to defeat Ieyasu at Mikatagahara in 1572; tradition says that Ieyasu fled from the field with just five men. Ieyasu, though a brilliant politician, was never a top-flight general, and he wisely decided to avoid fighting Shingen again and instead to adopt Fabian tactics. Fortunately for him – always luck rode with Ieyasu – Shingen was killed at a siege in 1573 and succeeded by a less able son. When Shingen's Takeda clan attacked Nagashimo castle in his home province of Mikawa, Ieyasu was finally able to persuade Nobunaga to take the events in the eastern realms seriously. This time the fearsome Nobunaga came in person, at the head of 30,000 men. Joined by Ieyasu's 8000 troops, the combined force smashed the Takeda at Nagashino on 28 June 1576. Although Shingen's son escaped and was able to carry on guerrilla warfare for a further six years, Nobunaga rewarded his loyal vassal by giving the Takeda province of Suruga to Ieyasu. It took the combined might of Nobunaga and Ieyasu until 1582 before Kai province was completely pacified and they were finally able to bring Shingen's son to bay. When he and his kinsmen were again heavily defeated at Temmokuzan in 1582, they committed seppuku (suicide), bringing the Kai wars to an end.

RISING STAR AND POLITICAL JUGGLER

By this time Ieyasu was already a rising star in Japanese politics and widely recognized as a major player; Nobunaga rated him among the top 12 damyos in the country. After 1573, when he threw out the old Ashikaga shogunate (which had lasted 230 years), Nobunaga was the real ruler of Japan – for the emperor was a mere figurehead – and Ieyasu's fortunes were closely bound up in his. He had navigated the treacherous currents skilfully, even surviving a scandal in 1579 when his wife and eldest son were found guilty of conspiring with the Takeda to assassinate Nobunaga. Ieyasu shrugged his shoulders, regarded the outcome as Buddhist karma and did not demur when Nobunaga had them executed; he simply nominated his third son, Hidetada, as his heir. But just when Ieyasu was about to ascend to an overt position as second man in the empire, Nobunaga was assassinated by Akechi Mitsuide. At the time of the killing Ieyasu was far from his home base, in the port city of Sakai, and had to make a secret and perilous journey overland to Mikawa province, evading Mitsuide's hit men who had identified him as another head of the hydra that was Nobunaga. Once back in Mikawa province he mustered an army for a war of revenge, only to find that Toyotomi Hideyoshi had beaten him to the punch and defeated Mitsuide at the battle of Yamakazi. Ieyasu recognized that Hideyoshi was now the power in the land, so left him to his conquest. But it was typical of him that he used Nobunaga's death to pick off provinces ruled by his vassals that he had been pledged not to interfere with while his patron lived; he promptly invaded the Kai region and added it to his portfolio.

Hideyoshi's rise to supreme power presented Ieyasu with his most difficult political juggling act to date. On the one hand, he knew he was not strong enough to oppose Hideyoshi openly, especially as the new hegemon showed that he was even more talented as a military captain than Nobunaga. On the other hand, it would be politically foolish and personally shameful to give in to Hideyoshi too easily. Ieyasu decided to see how far he could push without alienating Hideyoshi outright. He began by discreetly supporting Nobunaga's second son and heir against Hideyoshi's allies. Then he singed the great man's beard even more notably by fighting two limited engagements against Hideyoshi allies at Nagakute and Komaki as a kind of proof of toughness. But in 1582–3 Hideyoshi had his hands full elsewhere, contending for mastery against the other pretenders who would inherit Nobunaga's mantle, principally the leaders of the Shibata and Shizu clans. Ieyasu cunningly stayed neutral in this conflict, devoting himself instead to fending off the Hojo clan, who invaded Kai in pursuit of their own territorial claims. Ieyasu's army and the Hojo forces circled each other, but in the end made peace on a compromise basis: Ieyasu would retain Kai and Shinano provinces while the Hojo possessed Kazusa province. But the day of reckoning was soon to come, for Hideyoshi made short work of his principal enemy, Katsuie Shibata, and annihilated his forces at Shizugatake.

Hideyoshi was now unquestionably the number-one warlord in Japan, and Ieyasu's policy of supporting Nobunaga's son and heir began to look supremely perilous. At first Hideyoshi was inclined to punish this pestilential gadfly on his flank and sent a large army against Ieyasu. Long months of marching, feinting and counter-feinting ensued in the so-called campaign of Komaki before Ieyasu managed to win a pitched battle at Nagakute. Realizing that a supreme effort would be needed to dispose of Ieyasu, which would allow the ill-pacified elements in western Japan to rise up against him, Hideyoshi sensibly opted for a negotiated settlement. He confirmed Ieyasu as the ruler of the five provinces of Mikawa, Suruga, Kai, Totomi and Shinano and in return Ieyasu made formal submission as Hideyoshi's vassal at Osaka castle in 1586. To seal the bargain he gave his second son to Hideyoshi to be adopted and married one of his sisters to the great man; Hideyoshi's mother was sent to Ieyasu as a hostage. The two rulers regarded each other with a wary mutual respect, but Hideyoshi never completely trusted the powerful vassal on his eastern flank; the fact that Ieyasu made no contribution whatever to Hideyoshi's successful invasions of Shikoku and Kyushu in western Japan did not go unnoticed. Nevertheless, with Hideyoshi fully occupied in the west and Ieyasu trying to consolidate in the east, the alliance held up well for the next four years.

TOYOTOMI HIDEYOSHI — INNOVATOR AND ROLE MODEL

If Ieyasu had learnt much from Nobunaga, he became a truly wise man by observing Hideyoshi and copying his most successful innovations. Hideyoshi was in many ways an even more distinguished and talented person than Ieyasu, for he took Japan in new directions, whereas Ieyasu did little more than build on his achievement. Perhaps the most remarkable aspect of Hideyoshi's career was that he came from farming stock and gradually worked his way up the social ladder. As a young man he made a living from sewing, then entered the service of a rural samurai before leaving him to join Nobunaga's staff, where he rose to the top, eventually becoming his most talented general. Hideyoshi proved to be brilliant at siegecraft, brilliant at logistics and mass mobilization and brilliant at organizing the infrastructure of war, building roads, bridges and transport ships that could take his armies swiftly from one end of Japan to the other. His talent also manifested itself in military innovation. He built the first armour-plated ships with which he eventually annihilated the Mori fleet, consisting of wooden ships. And he was the first to use harquebuses to devastating effect, using three ranks of musketeers firing in rotation to win the battle of Nagashimo in 1573 and anni-hilate the Takeda clan's mounted warriors.

We have spoken earlier of the 'centralized feudalism' of late six-teenth-century Japan – an amalgam of the 'early modern' societies of western Europe and the late entry of the Nipponese nation into feu-dalism. If any one person can be described as being responsible for introducing this system, it was Hideyoshi. First he defeated his most powerful rivals and managed to unite all the warring damyos into a federation; Nobunaga had gone some way with this, but he was more of a ruthless destroyer than a nation builder or a creator of enduring institutions. The feudal nature of Hideyoshi's Japan was illustrated by the way the damyos surrendered their lands to Hideyoshi and received them back as fiefs after swearing allegiance to him. Meanwhile, nearly all rural samurai were now concentrated in the castles of the territorial damyos. Hideyoshi then kept village society on the edge, unable to plan rural rebellion, by taxing two-thirds of their rice crop as land-tax; financially astute, along with all his other qualities, Hideyoshi was even able to manipulate the mar-ket price of rice. It was only under Hideyoshi that feudalism proper came in, for he shifted the agricultural base from slavery to serfdom. In the towns he proved far more successful than Nobunaga in assert-ing his authority over the merchant guilds and other urban power nexus. In a decree of 1585 he finally abolished the guilds, allowing him personal control of economic life. By smashing the power of the villages and forcing all samurai to live in castles with the damyos, he broke the link between samurai and peasantry. Henceforth, he needed only to control the powerful territorial damyos to control everything.

Hideyoshi also ordered the Japanese equivalent of a 'Domesday Book', attempting cadastral surveys that would give him a complete picture of the nation's social and economic life. He announced a peace policy that would bring tranquillity to all the islands of Japan, banned piracy in a decree of 1588 and passed a law confiscating all weapons held by farmers. He then ingeniously used the weapons so confiscated for religious purposes by melting them down to build the great temple of Buddha in Kyoto – an inspired use of religion for social control. Local damyos were ordered to keep a strict surveillance on all who made a living from the sea and crucify anyone taking to piracy; this was one way of making sure that farmers, now deprived of the village samurai, could never make common military cause with fishermen.

A brilliant politician, when waging war Hideyoshi used mixed force drawn from the contingents of various damyos, so that there could be no mass desertion or treachery; he made the troops from the most recently allied damyos fight in the vanguard to prove their loyalty. Having abolished the merchants' guilds, he flattered and cajoled the most powerful of them, with the result that trade blossomed in his time. He transformed the structure of trade and put the currency on a sound footing, mainly by dramatically increasing the production of bullion; in his era Japan produced one-third of the world's gold and silver. He was also the first powerful lord to become alarmed at the rising power of the Portuguese in their enclave at Nagasaki and to have misgivings about the activities of the Jesuits, who had flooded in after Francis Xavier made his famous missionary journey in the 1540s. Hideyoshi's centralized state introduced the idea of *kogi*, his own authority, superior to that of all other damyos and even the imperial system itself. Nobunaga had shown the way by abolishing the compulsory labour peasants performed for their local lords and decreeing that the only legal corvée was the one he himself ordered. Hideyoshi's 1587 decree went one better: it stipulated that whereas vassals could be moved from one fief to another, peasants and farmers were tied to the soil.

THE POWER OF PERSUASION

Most of all, Ieyasu learnt from Hideyoshi that it was always better to conciliate than to coerce, to regard warfare as the last resort. Hideyoshi liked to win over reluctant allies by persuading them that they would do better as his vassals than as independents. Ieyasu admired the way Hideyoshi gradually eliminated or neutralized all rivals and how he silenced the snobbish Japanese oligarchs, blindly obsessed with the minutiae of rank and lineage. Knowing of the whispering about his 'low' origins, Hideyoshi had himself adopted into the Fujiwara family, hereditary custodians of the office of regent – a post Hideyoshi also took for himself. Finally, he changed his surname to Toyotomi and, as there was no longer a shogun, Hideyoshi now had the name of it as well as the fact. When he defeated the recalcitrant Shimazu clan of Kyushu in 1587 and their allies the Chosokabe in Shikoku, that left just four super-damyos for him to manipulate like a puppet master, and one of these was his hyper-loyal vassal Ieyasu. Hideyoshi decided that the next piece to be taken off the chess-board should be the Hojo of the Kanto – eight provinces in northeastern Japan. Ieyasu had cleverly conserved his military resources while Hideyoshi campaigned in the west and south of Japan, using his power instead to order land surveys and get the land tax paid in rice – both innovations he owed to Hideyoshi. But when Hideyoshi declared war on the Hojo, Ieyasu could no longer sit on the fence, even though Ujimasa, damyo of the Kanto, was his friend. When Hideyoshi led a huge army of 160,000 against the Kanto, Ieyasu had to add his own 30,000 troops to the invading soldiers.

This gigantic combined assault force made straight for the Hojo headquarters at Odawara castle. Hideyoshi made sure that this time Ieyasu did not evade his obligations and used his 30,000 men as the front-liners in a bitter six-month siege in 1590, which ended with the mass suicide of the Hojo leaders, thus ending 450 years of this clan's dominance in the north. Once the Odawara castle fell and the Hojo were defeated, the few remaining nominally independent damyos in the north, especially the Date clan, made their submission. Hideyoshi was now master of all Japan and, once again, he proved himself a superb machiavellian. Thinking that Ieyasu was by now too powerful, he ordered him, as his vassal, to change fiefs; instead of the five provinces he currently ruled, including Mikawa (the others were Totomi, Suruga, Kai and Shinano), he was to take over in the Kanto. Superficially a promotion, as possession of the Kanto would make Ieyasu the greatest landowner in the country, greater even than Hideyoshi, the transfer was construed by Ieyasu's followers as a disguised exile. It uprooted him from his power base and placed him right on the periphery, in Edo castle, far from the centre of events. The mass exodus of thousands of families was accomplished with astonishing speed, and 18 days after leaving Mikawa Ieyasu and his court were ensconced at Edo. Hideyoshi made clear how his mind was working by installing his most trusted vassals as lords of the lands vacated by Ieyasu.

IEYASU'S FINEST HOUR: VICTORY THROUGH CUNNING

In some ways this was Ieyasu's finest hour, a real test of character and moral fibre. A lesser man would have taken the hint and been content to be a peripheral back number but Ieyasu brought his usual energy to his new domains. The Kanto remained untouched by Hideyoshi's feudal system, so that there were plenty of recalcitrant village samurai at large – and it was for precisely this reason that Hideyoshi had made the transfer, thinking he was placing his mighty vassal on a bed of nails. Ieyasu accepted the challenge and, in a remarkably short time, imposed the Hideyoshi system he had himself adopted in his old fiefs of Mikawa. Or rather, he did so partially, rejecting a one-size-fits-all solution and steering away from trouble if the introduction of a monolithic feudalism meant local disruption. Amazingly, he soon won over most of the sullen and disaffected Hojo samurai who were mourning the demise of their clan, and transformed them with lands and bureaucratic places. Soon Hideyoshi was asking himself if he had made the wrong decision, for Ieyasu even improved agricultural productivity; too late Hideyoshi realized that he had underrated the economic potential of the Kanto, that in fact gross production could soon amount to three times what Ieyasu had generated in his original fief. Ieyasu showed great political skill in the way he partly tightened his control over the Kanto by expanding his bureaucratic and administrative personnel, and partly bought off the opposition. In a Solomonic parcelling out of lands and wealth, he took one-third for himself, gave his senior damyos one-third and left another third for the lesser vassals. Since the Kanto was a political backwater, Ieyasu enjoyed almost total local autonomy and was now unquestionably the number-two power in Japan. The move to the Kanto had paid off handsomely: as one writer later commented: 'Ieyasu won the empire by retreating.'

As long as Ieyasu made no move to enter the arena of national politics, Hideyoshi was content to leave him in splendid isolation in his eastern domains. More concerned about the morose and sullen conquered clans of the west, Hideyoshi decided to displace their latent martial energies against an external foe – one of the classic ploys of autocrats throughout history. In 1592 he sent a huge invasion force (158,000 troops) to Korea – the prelude to a planned later assault against the Ming dynasty in China. In a bizarre prequel to another Korean war 350 years later, the Japanese armies at first swept all before them, benefiting from the experience of decades of domestic warfare and their deadly musketeers, but then the Chinese crossed the Yalu river in support of beleaguered Korea. The war soon bogged down into stalemate, and Hideyoshi lost face as a result. Ieyasu kept out of the conflict and sent no troops to the theatre of war – a repeat of his ploy when Hideyoshi was conquering the western clans. Perhaps thinking he could not trust Ieyasu to remain quiescent in the east while so many of his own troops were in Korea, in 1593 Hideyoshi summoned Ieyasu to take up permanent residence in Kyoto as his military adviser. This was a typical Hideyoshi tactic. When he had conquered a clan, he always ordered that the wives and children of the defeated damyo had to live with him as hostages in the capital city. He also required the quasi-permanent presence of the damyos themselves, and they were granted holdings to pay for and maintain residences in Kyoto, much as if they were ambassadors from a foreign court; obviously it was very difficult to raise the standard of revolt if you were under virtual house arrest. In some ways Ieyasu could count himself lucky that Hideyoshi had not required his presence before.

AMID KYOTO'S POLITICAL TURBULENCE

While he was in Kyoto, Ieyasu's sons and loyal retainers controlled Edo and the Kanto for him. We do not know what role he played in the turbulent politics of the 1590s, caused by Hideyoshi's faltering political touch. The first sign that the great man was no longer the penetrating mind of old came in 1591 when he retired from the regency – the position he officially held as the real power in Japan – and handed over to his adopted son, Hidetsugu, but without relinquishing overall control. The sequel was predictable. Soon Hidetsugu's decrees were clashing with his father's; it became apparent that the one flaw in Hideyoshi's mental apparatus was that he did not understand the law of self-sustaining bureaucracies and failed to see that the court system (the regent used civil servants whose primary loyalty was officially to the emperor) had a momentum of its own, at least if not controlled by a strong hand. That hand was not Hidetsugu, and it is clear that Hideyoshi should have dismissed him and ruled by diktat. Instead he let matters slide and by 1594 tension between his followers and those of Hidetsugu was clearly visible. Suddenly Hideyoshi accused Hidetsugu of treason and ordered him to commit seppuku; finally Hideyoshi threw off the mask and emerged as absolute and sole ruler.

It is not certain if it were by Ieyasu's advice that a truce was agreed with the Chinese for four years (1592–6), though it would have been in his interest for the war to drag on, wasting Hideyoshi's blood and treasure. Whatever devious plans Ieyasu had in mind, he received an unexpected shock in 1593 when Hideyoshi sired a son, Toyotomi. These are the obscure years in Ieyasu's biography, but there are signs that he was already laying plans against the time of the succession, for the ageing Hideyoshi suddenly seemed to be further losing his grip. In 1597, losing patience with the protracted peace negotiations with the Chinese, he ordered a second invasion of Korea; this time an army of 140,000 troops departed the shores of Japan. The Chinese and Koreans opposed the invaders with guerrilla tactics, and soon food supplies for the Japanese soldiers began to run short. Korea was a smoking ruin of devastated and deserted villages, rice crops were no longer harvested, and it was said that a crow flying over central Korea would have to carry its own provisions with it. Increasingly exasperated by a war he seemed unable to win whatever he did, Hideyoshi once again opened peace negotiations but died soon afterwards (18 August 1598). Before he died he left behind a shirt of Nessus in the form of a five-man Council of Regency. The other four notables (apart from Ieyasu) were Maeda Toshiie, Mori Terumoto, Ukita Hideie and Uesugi Kagekatsu, all important names in Ieyasu's biography.

The route to power

Ieyasu's political skills were now taxed to the limit, for the other regents were bent on a continuance of the Toyotomi dynasty and system. In addition, the five regents themselves were coming under severe challenge as decision-makers from another of Hideyoshi's institutions, the Five Commissioners, supposedly only administrators but inevitably rivals for supremacy with the regents. As Japan threatened to descend into chaos, the great damyos of Nippon faced an obvious choice: either they should back the regents and declare for the succession of Toyotomi or they should come out for Ieyasu. Many Toyotomi damyo – those at least who had freedom of action – decided that Ieyasu would be the winner in any nationwide trial of strength and went over to him. But the western damyos were determined that they would never be ruled from Edo and, besides, there were many great lords whose fortunes were entirely tied to Hideyoshi and his successors and who would lose heavily if Ieyasu became shogun. The one advantage that Ieyasu clearly did enjoy was that the opposition to him was bedevilled by factionalism. In an uncanny echo of the élite divisions in old Byzantium, the anti-Ieyasu notables were split between a bureaucratic faction represented by the five commissioners based at Osaka and the military lords of western Japan. This was a wound that would never heal and would contribute materially to Ieyasu's eventual triumph.

Immediately on Hideyoshi's death, Ieyasu raised the stakes by moving into Fushimi castle, the late ruler's personal fortress. The other regents regarded this as usurpation but drew back from pitching the nation into civil war. At first they made half-hearted attempts to assassinate Ieyasu, but none of the plots worked.

In 1599 the most senior regent, Maeda Oshiee, died, removing a principal obstacle to Ieyasu's route to power; Oshiee had immense prestige and gravitas, and it would have been hard for Ieyasu to defy him openly. Once he was gone, Ieyasu began issuing decrees as if he already were Hideyoshi's successor. Understandably irate, the other three regents prepared for war and called in the western lords and the commissioners. Gradually the prime mover and nerve centre in the opposition to Ieyasu emerged more clearly as Ishida Mitsunari, one of the five commissioners, a 'fixer' and broker of deals but not a soldier. Under Ishida's aegis a coalition of western septs gradually formed, with the main contingents being supplied by the Date, Mogami, Satake and Meda clans. But Ishida was a disastrous choice as coordinator of the western allies. Originally promoted by Hideyoshi merely for his skill in the traditional tea ceremony, he was an intriguer with a brusque manner who alienated many by his lack of charm; some clan leaders opted to fight for Ieyasu for this personal reason alone. The great military talent among the western-ers was Mori Terumoto, head of the powerful Mori clan, but the jealous Ishida, who had no martial abilities himself, refused to give Terumoto his head; in a classic tail-wagging-the dog situation, the bureaucrat Ishida, who had scarcely heard a shot fired in anger, ended up directing a complex campaign while his most able general languished on the sidelines.

In the coming trial of strength between the lords of western and eastern Japan, Ieyasu, the lord of the east, held most of the cards. As ruler of the Kanto, Japan's breadbasket (or rather ricebowl), he commanded domains worth 2.5 million koku annually (the 'koku' was a traditional Japanese measure of wealth, equivalent to 180 litres/380 pints of rice – the amount believed necessary to keep a man alive for a year). His long years of patient planning and intrigue had paid off, to the point where wiseacres quipped that Nobunaga was the man who had kneaded the rice cakes, Hideyoshi the man who cooked them, but Ieyasu was the one who ate them. Undoubtedly, his most dangerous opponent was Mori Terumoto, also a wealthy man, with personal holdings of 1.2 million koku annually. It was said that

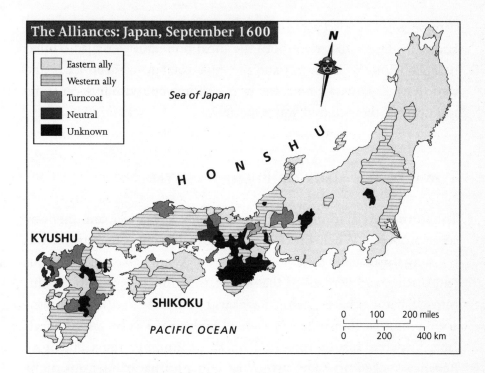

whereas Ieyasu could build a road from the Kanto plain to Kyoto, Terumoto could construct a bridge of gold from his domains to the capital. Yet beyond this, Ieyasu and his allies fought for a clear objective: to see the lord of the Kanto enthroned as shogun; discipline and unity characterized his forces and allies. The western clans had no such clarity of aim, except a vague commitment to keeping Hideyoshi's legacy and his dynasty in being.

Suspicious of each other's motives, and with many suspecting that the ambitious Ishida ultimately aimed at supreme power for himself, they were far more vulnerable to having clans and damyos suborned by Ieyasu. The first such shock was the open defection of Kato Kiyomasa, a veteran of the Korean campaign and a Hideyoshi loyalist, simply and solely because of his hatred of Ishida. Kiyomasa was to play a crucial role in the coming campaign, as he prevented Kyushu from joining the western alliance. Moreover, the western damyos were divided against each other on ideological and religious issues. In their ranks were both the most powerful Christian damyo in Japan (Konishi Yukinaga) and the most vociferously anti-Christian lord (Kato Kiyomasa), both veterans of Korea but with little in common except their shared war experiences.

TOWARDS THE SIEGE OF FUSHIMI CASTLE

The formal declaration of war awaited only a *casus belli*, and this was soon provided. When the western damyo Uesugi Kagekatsu began building up an army to oppose the lord of Kanto, Ieyasu asked for an explanation and demanded that Kagekatsu come to Kyoto to explain himself before the emperor. Kagekatsu responded with a counter-accusation that Ieyasu had broken all Hideyoshi's rules and mocked Japan's sacred institutions. At Ishida's prompting, the Council of Regency backed up Kagekatsu by issuing a formal condemnation of Ieyasu, indicting him on 13 counts, including the contracting of unilateral political marriages and the occupation of Hideyoshi's castle. While Ieyasu was heading north to chastise Kagekatsu, Ishida announced the formation of a second army, seized a number of Tokugawa retainers as hostages and imprisoned them in Osaka castle. Ieyasu had to leave his subordinates to deal with the Uesugi troubles while he marched west against this new threat. Meanwhile, Ishida lost face disastrously with a scandalous episode at Osaka castle. Among the Tokugawa hostages he had taken was a highly educated Christian lady named Gracia Tadaoki, the wife of one of Ieyasu's generals. Determined that she should not be a 'burden' either to her husband or his feudal lord, she ordered a servant to kill her – doubtless to get round the Christian veto on suicide – and then set fire to her quarters. In the confusion following this sensation, many of the other hostage women managed to escape. Ishida lost caste on two grounds: both that he had not placed sufficient guards to prevent the lady effectively committing suicide, and that the incident cast doubts on his general judgement. If he could not control Osaka castle, how could he deal with the wily and resourceful Ieyasu?

Ieyasu meanwhile followed his usual practice of considering actual fighting merely the last resort when all other measures had failed. His first ploy was to try to detach as many of the wavering and dubious allies of the western alliance by bribes and lavish promises of future preferment. His greatest success was with Kobayakawa Hideaki, the 19-year-old nephew of one of Hideyoshi's wives and the late ruler's foster son. Like so many of those in the western alliance, Hideaki was brooding about an insult from the hot-tempered and indiscreet Ishida. During the Korean war Ishida denounced Hideaki to Hideyoshi as a callow incompetent, as a result of which the young man was recalled from the front in disgrace. Hideaki remembered that it was Ieyasu who had interceded for him with Hideyoshi and got him reconciled to his foster father. When Ieyasu approached him and offered him the position of 'kanpaku' (chancellor) in the government he hoped to form when he had defeated the westerners, Hideaki had a double motive to support the east, both a debt of honour and the hope of future preferment. Everything pointed to an entente with Ieyasu, but first Hideaki consulted his stepmother (Hideyoshi's widow) about what he should do; she recommended following his conscience. Correctly reading this as a euphemism for 'self-interest', Hideaki told Ieyasu that what he asked was difficult but he would do it; Ieyasu saw that the maximum purchase to be gained from this new recruit was to allow him to continue as an ostensible member of the western alliance, ready to apostatize when the time was ripe.

Further probing by Ieyasu's secret agents, especially among the Mori clan, indicated that there were other potential turncoats who would certainly not give their all in a climactic battle; to them also Ieyasu made lavish promises and left them to brood over the potential rewards of treachery. Another valuable secret recruit was the damyo Kikkawa Hiroie, yet another motivated by personal pique towards Ishida. Feeling that Ishida's insistence on directing operations was an insult to Terumoto and the Mori clan, Hiroie sent clandestine assurances to Ieyasu that when the time came his men too would not fight, and he hoped to encourage the entire Mori clan to act likewise. Ieyasu's spies even brought back encouraging news from Kato Kiyomasa, the violently anti-Christian damyo. By promising that, if successful, he would root out the Christian 'cancer' in Japan, Ieyasu dangled a bait that made Kiyomasa think twice and thrice.

Ishida and his allies temporarily withdrew from Osaka castle when Ieyasu marched west, hoping to stretch his lines of communi-

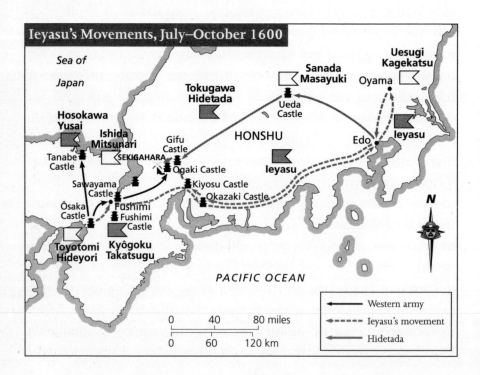

cations and lure him westward to his doom. Ieyasu spotted the trap and marched slowly back to Edo, leaving on 24 July 1600. Next evening he stopped at Fushimi castle and spent the evening drinking with the keeper, an old friend named Torii Mototada. It was clear to both that the western army would attack the castle and overwhelm Mototada and his samurai, so that the carousing was more in the nature of a sad farewell; the pathos in the parting next morning later became a favourite subject for Japanese poets. Ieyasu then proceeded to Edo (reached on 10 August) where he remained until 1 October, building up his army to the right pitch and making sure all commissariat and logistical problems were properly attended to. Ishida meanwhile called a council of war at Sawayama castle on 17 August, where all the powerful western lords attended, including the regent Ukita Hideie and the secret turncoat Kobayakawa Hideaki. As predicted, they made Fushimi castle their first target and began a ten-day siege on 27 August. Mototada and his samurai performed valiantly and held out until 6 September, but in the end one of the towers was betrayed by a fifth columnist and Ishida's men swarmed in; Mototada and his remaining men committed seppuku. Mototada had more than fulfilled his pledge to Ieyasu: he had held up the western advance and cost Ishida 3000 casualties.

Towards a decisive clash

It was clear that the decisive clash between east and west would take place close to modern Nagoya, where Honshu, the main island of Japan, narrows to a waist between the Pacific Ocean to the south and the Sea of Japan to the north. The western allies mustered at Ogaki castle, apparently preparatory to an invasion of Mikawa. But Ieyasu pre-empted them by sending large forces to take the two castles of Kiyosu and Gifu, on either side of Ogaki and about 24 kilometres

(15 miles) distant; whoever controlled these fortresses would also secure the Tokaido and Nakasendo roads, enabling armies to march directly on Osaka and Kyoto. Kiyosu was a walkover, as it was held by a grandson of Nobunaga, who declared for Ieyasu, but Gifu, held by western sympathizers, did not fare much better and fell on 28 September. When word reached Ieyasu in Edo, he set out west with his main army, hoping to force a decisive conclusion. To make sure that there was no attempt at a stab in the back while he marched westward, he sent his third son and heir apparent, Hidetada, on a northwesterly sweep through Honshu, with instructions to meet up with him at a rendezvous point on the Gifu/Kiyosu line. Hidetada, intoxicated by the joys of independent command, proved as disastrous to his father as Marshal Grouchy would to Napoleon at Waterloo. Lusting for martial glory, he thought he would knock out Ueda castle in central Honshu and then make his father an unexpected present when he met him. But Hitetada, with 36,000 men, got bogged down in a protracted siege of Ueda. Hoping every day that he could complete the siege and still reach his father in time for the final battle, he tarried too long, and when he finally broke off the abortive siege, he found he had left himself too far too march: he never made the rendezvous and never fought alongside Ieyasu in battle.

As if to compensate, Ieyasu's defenders at the simultaneous siege of Otsu castle by the western army held Ishida's men at bay; only 3000 ranged against 15,000 attackers, they defended fanatically and, when the fortress finally fell on 21 October, it was already too late for the western forces to play any part in the climactic battle. The grim and bloody struggle for Otsu (on the shores of lake Biwa), which at one stage saw the involvement of Japan's famous professional assassins, the Ninja, turned into something of a spectator sport when onlookers were able to view the conflict from nearby hillsides and even the rooftops of Kyoto; this was a pre-echo of the famous picnicking day-trip at Bull Run in the American Civil War (1861), when aristocratic ladies similarly 'enjoyed' a day's bloodshed.

Meanwhile, Ieyasu continued to advance rapidly from the east. Ishida, alarmed at the speed of the easterners' approach, began to fear he would be outnumbered and sent out frantic pleas to the Mori clan, who finally brought their 30,000-strong army to Osaka. Yet the initiative still lay with Ieyasu. When he reached Kiyosu on 17 October he had the option of bypassing the western army in Ogaki and marching straight to Osaka or Kyoto on the trunk roads he now controlled. But this seemed to him merely like postponing the inevitable; better to go for the jugular now, before the west could increase in strength. Buoyed by a further message from the turncoat Kobayakawa Hideaki that he would definitely switch sides once battle was joined, Ieyasu made straight for Ogaki, arriving in the environs (at the village of Akasaka, just 5 kilometres/3 miles away) on 20 October. Camped by a river, he spent the afternoon and evening beating off heavy skirmishing parties from Ishida, who tried to take the bridge linking Akasaka with Ogaki.

SEKIGAHARA: THE CONCEPT OF THE CRANE'S WINGS

This was the time for imagination by the west and Shimazu Yoshiro, a talented general, suggested the way forward: a full-scale night attack on Ieyasu before he had drawn up his forces properly; the regent Ukita Hideie, second in command, enthusiastically supported the idea. Yoshiro argued that Ieyasu's forces had been marching non-stop for nearly two weeks, eating and sleeping in their armour, and so were tired and demoralized. Almost predictably, Ishida and his advisers rejected the proposal, arguing (incorrectly) that numbers were on their side and that night attacks were the desperate expedient of weak forces attacking stronger ones. Yoshiro was gravely insulted by the rejection of his advice and the implicit criticism that it was cowardly and somehow dishonourable, so here was yet another powerful damyo resentful of Ishida and brooding about contemptuous treatment.

Ishida made it clear to Ieyasu that he was accepting his challenge by moving his forces to the plain of Sekigahara, a natural battlefield almost completely surrounded by mountains. His army set out to trudge to Sekigahara in driving rain, sleety precipitation blown nearly horizontal by the wind. The advance to the battlefield on such a cold, dark, rainy, windswept night was scarcely an uplifting experience: unable to see very far in the gloom, the vanguard of some regiments kept bumping into the rear of others, setting off momentary panic that the enemy was being encountered. Reaching the plain around 1 a.m. on the morning of 21 October, Ishida at once consulted with Hideaki, who had taken up position with his 15,000-strong force on Mount Matsuo to the south of the battlefield. Ishida extracted from Hideaki, about whom he may have had some doubts, a solemn promise that his troops would descend on Ieyasu's flank as soon as he lit a signal fire.

Ishida's strategy was textbook and, but for internal treachery, might well have been successful. It was the classic crane's-wings approach described in ancient Chinese war texts, where the bird's wings eventually closed over the prey once it had swooped. The intention was to catch the advancing Ieyasu in a kind of human box canyon, with Mount Matsuo to the south, Mount Tengu to the east and Mount Sasao to the north. As Ieyasu entered the trap, the Mori clan would work round behind him from the south and come in on his rear, thus completing the encirclement. The main forces under Ishida and Ukita Hideie would stand firm and breast the initial onslaught from the eastern army. Then, at precisely the right moment, Ishida would light the beacons, the Mori would seal off the eastward retreat and Hideaki would charge down on the enemy's left flank from the heights of Mount Matsuo. No good general would have ventured into such a cauldron – the risks were too great – but Ieyasu had an ace in the hole that Ishida knew nothing of. Not only was Hideaki secretly pledged to him, so that the flank attack would never take place but, even worse, in the vanguard of the Mori 'seal-off' regiments, which were meant to work around behind Ieyasu, was none other than Kikkawa Hiroie, self-styled hero of the Mori clan, still fuming at the insult offered to Mori Terumoto and determined to pay out the hated Ishida.

Ieyasu's army got under way about 2 a.m., the troops trekking through heavy mud. Once in the valley, he set up his command post at the bottom of Mount Nangu, in a rude headquarters consisting of several bamboo poles ringed with camp curtains and a rough mat on the floor. Both armies now dozed fitfully in their rain-soaked armour, waiting for daybreak. As dawn broke through falteringly, the rain stopped and turned into a heavy mist. The journal from Ieyasu's personal physician survives and reads as follows: 'Slight rain. Dense fog in the mountain valley. Can't see ninety yards. Fog lifted briefly and could see two or three hundred yards; fog then grew dense again. Barely made out enemy banners. On horseback, Lord Ieyasu made out positions of Ishida Mitsunari, Konishi Yukinaga and Otani Yoshitsugu. Estimate distance at two and a half miles.'

BATTLE DISPLAY ON SEKIGAHARA PLAIN

Some 170,000 troops were crowded into the plain of Sekigahara on the memorable day of 21 October 1600. The eastern army of Ieyasu was slightly larger, about 89,000 strong as against the western alliance's 82,000. The best units on either side were, respectively, Ieyasu's own 30,000 samurai and the 17,000 under the command of the regent Ukita Hideie. Amid the wheeling of troops and the neighing of horses, the senses would most likely have been engaged by the riotous colours, the variety of armour and the positively totemic assortment of headgear. Most Japanese harquebusiers wore a solidly riveted cuirass of steel plates, but the simple infantryman had either a basic belly-plate or a folding armour of hexagonal plates and some form of helmet. Additionally they wore padded sleeves with integral splints of lacquered metal and mail sewn on and some rudimentary leg armour, both greaves and thigh protectors. Their principal weapons were a long lance and two different kinds of sword, with blades from a foot to two feet in length. Élite units, their leaders and the damyos wore special, elaborate armour and intricate helmets. At Sekigahara, Ieyasu wore a special suit of European armour, presented to him by the Portuguese, although he disdained a helmet, wearing a conical hat instead. Perhaps he wished to preach the virtues of simplicity and contrast himself with Hideyoshi, who was famous for his sunburst headgear.

The warlord Ii Naomasa, who fought with Ieyasu, commanded a 3600-strong division called the Red Devils, all clad in red-lacquered armour; these had a ferocious reputation, like that of the Viking berserkers. The most striking insignia were the helmets and banners of the damyo. Ii Naomasa and his men had two gilt horns protruding from the top of the helmet, much like an early television aerial. Date Masasuma wore what looked like a stag's head on his helm while Honda Tagatsuku wore a kind of fitted crown that looked like a decrescent moon. The headdresses were surely the most bizarre and elaborate seen in advanced warfare since the days of the Vikings; in the sixteenth century possibly only the Aztecs could top the Japanese when it came to head plumage.

Adding to the sensuous confusion was the plethora of banners. The 'sashimono' was a long banner worn on a crossbar and pole device attached to the back of the armour and representing the crest

An aztec codex portraying human sacrifice: the ritual that so horrified the conquìstadores.

A mansucript detail of the conquest of Mexico: a violent collision of cultures.

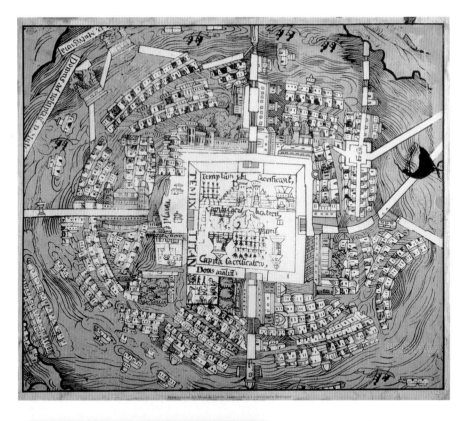

This map, attributed to Hernando Cortés, depicts the labyrinth that was the Aztec city of Tenochtitlan. The Spanish were lucky to escape from it and even luckier to reconquer it.

Cortés: the portrait shows his ruthlessness and determination. This is a man who would even have defied the king of Spain himself if he had not fallen in with his wishes.

Lacrimae rerum: the twilight of the Mexica.

Hollywood's version of the meeting of conquistador and Aztec. The 1947 movie *Captains from Castile* treated the conquest of Mexico almost as a footnote in its own interpersonal drama.

Hanging scroll painted with an image of Tokugawa Ieyasu: a corpulent genius. As Churchill later showed, obesity is no barrier to supreme political leadership.

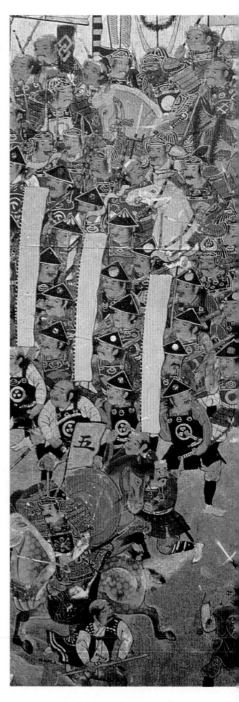

Detail of part of a folding screen that depicts Sekigahara where at least 50,000 perished on 21 October 1600. Following Bushido custom, Ieyasu had 40,000 skulls of the enemy fallen meticulously counted.

TOKUGAWA IEYASU

Japanese samurai
armour. The demonic
aspect was meant to
terrify enemy infantry.

Ieyasu's cavalry delivers
the *coup de grâce*. A
generic still from Akira
Kurosawa's 1980
masterpiece *Kagemusha*.

The young Napoleon according to the painter David.

Amphibious operations dominated the three-month struggle for Toulon. The French were superior on land and the British at sea.

Napoleon, the master of artillery. He not only identified the enemy's weak spot but grasped intuitively that massive artillery barrages would bring him victory.

Engraving showing the destruction of the French fleet at Toulon in December 1793. Some 11 ships were set on fire but only six were gutted. Incredibly, the Royal Navy had laid no contingency plans for the destruction of the French ships. In France, the incident caused great bitterness, which the British compounded by a similar destruction of French warships at Oran, Algeria, in 1940.

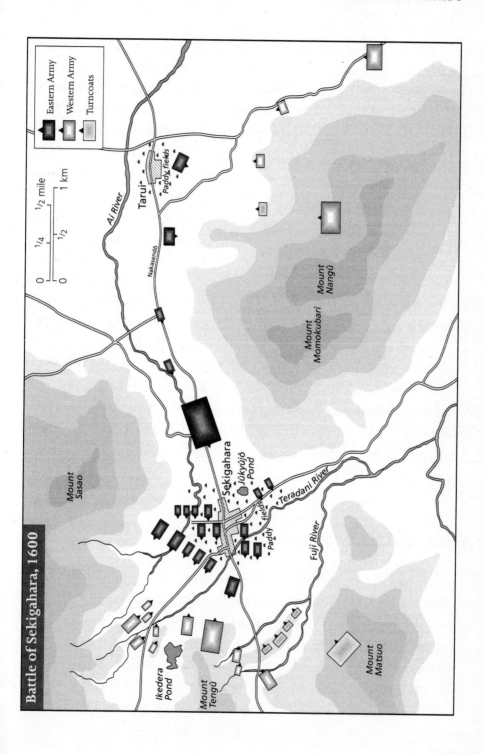

Battle of Sekigahara, 1600

Eastern Army
Western Army
Turncoats

¹/₂ mile
1 km
¹/₄
¹/₂
¹/₂
0
0

Ai River

Tarui

Paddy fields

Nakasendō

Mount Sasao

Sekigahara

Jōkyūjō Pond

Paddy fields

Teradani River

Fuji River

Mount Momokubari

Mount Nangū

Ikedera Pond

Mount Tengū

Mount Matsuo

of the unit and its commander. The damyos and lords had standard bearers to carry larger versions of these crests. Such emblems were supposed to be similar in shape, form and pictogram to those of other allies so they could be recognized in an instant and to be as unlike the enemy's as possible but, naturally, this was a doctrine of perfection and mistakes occurred, especially it was not unknown for an unscrupulous enemy to appear in the field at a vital moment sporting false colours. But the banners were important as rallying points, for the hierarchies of modern armies were almost unknown: feudal levies and samurai followed their damyos, and battalions, regiments, brigades and corps as we would understand them did not exist. Rivalries among the different damyos meant that overall discipline was difficult to maintain; in this respect the Japanese armies of the civil war period resembled the crusader armies of Richard the Lionheart and Philip of France. The most notable difference between the armies fighting at Sekigahara and those that contended fifty years earlier was the almost total absence of archers. The bow and arrow had been supplanted by the harquebus and the matchlock, generally considered the battle-winning weapon. Called 'teppo tai', these matchlocks were now mass-produced domestically from models introduced by the Portuguese.

Into battle with the Red Devils

Ieyasu had drawn up his men in three ranks, with his 30,000 samurai as the reserve. He had given the honour of leading the attack to Fukushima Masanori's 6000 men on the left wing of the first row, but even in the more disciplined eastern army, personal rivalries, prestige and the cult of 'honour' and 'face' weighed more than precise instructions from the supreme leader. Ii Naomasa took the view that he was a long-time Tokugawa loyalist whereas Fukushima

Masanori was a johnny-come-lately whose primary loyalty had always been to Hideyoshi; he was only with the easterners now because of his hatred of Ishida. He was therefore not going to concede pride of place to Masanori. On the pretence that he was conducting an inspection, shortly after 8 a.m. he rode out in front with 30 of his Red Devils.

It was estimated that the gap between the two armies was the distance a fit man could run in two minutes, and in a trice Naomasa's men were closing the range, a veritable red apparition, clad from head to foot in flame-red armour, with red banners, and even their lances painted red. Furious at this perceived impertinence, Masanori ordered his men forward to close with the opposing units, which were the élite troops of the regent Ukita Hideie on the right wing of the western army. While Masanori's musketeers poured a withering fire on the enemy, slightly to the south the 8000-strong far left of the eastern army crashed into a suspected gap on Ishida's right, while the right and centre of Ieyasu's army, 20,000 strong, made a mass attack on Ishida's command post. Ferocious hand-to-hand fighting ensued, with the easterners appearing to have the edge.

On Ishida's extreme left, Kuroda Nagamasa on the eastern side was thirsting to get at Ishida's reserves to avenge the insult sustained in Korea. But the Ishida men defended their position well and, to prevent collapse in this sector, Ieyasu was forced to order up more and more Togukawa gunners to pour a deadly fire into the flank of Ishida's front line. Once again Ishida responded well, by bringing up five heavy cannon normally used in siegework and blasting great fusillades at the enemy. The tactic worked, and the easterners fell back alarmed and chastened; however, they rallied sufficiently to beat off a counter-attack ordered by Ishida to try to seize the psychological advantage. Now Ieyasu spotted western forces trying to work around his rear and seal off the exit to the east and ordered a charge to push them back. The commander of the flanking force sent word to the 15,000-strong Mori division on Mount Nangu to come down and support them but, before the Mori could make a move, they had to receive the all-clear from Kikkawa Hiroie, who was nearest to the enemy and had plain sight of them. In danger of being cut off and surrounded, Ieyasu counted on treachery to win him the day, and, sure enough, his hopes were well founded. The Mori waited for the signal from Hiroie, but it never came. And meanwhile the press of battle was as hot as ever in the central sector: the dull crackle of matchlocks counterpointed the screams of the wounded and dying in the cavernous valley, producing the effects of a nightmarish echo chamber. To compound the hellish atmosphere, the black smoke from thousands of firearms made day seem like night.

A CRITICAL MOMENT DENIED

By 10 a.m. the battle had been raging for two hours with no clear overall advantage to either side. In three distinct locations there was a ferocious slugging match, with weary samurai hewing and slashing and the pop-pop of musketry ubiquitous. One of these mini-battle-fields was on Ieyasu's far left, another in the centre, but the most desperate encounter of all was between the right and the centre, where the regent Ukita Hideie's 17,000 picked troops were engaged in vicious, no-quarter combat with Ieyasu's centre right. The noise of battle rolled all over the plain as the struggle surged back and forth, with first one side, then the other, having the edge. Ieyasu moved his command post forward to within 1 kilometre (half a mile) of Ishida's banner, but his aides noted that for the first time he seemed uncertain of the outcome, tense and nervous when command decisions were put to him, flying into rages over trivia and biting his nails compulsively, as he always did when under stress. But the long-term effects of Ishida's abrasive personality were gradually being felt.

Shimazu Yoshiro had not forgotten the insulting and contemptuous way Ishida had dismissed his advice to make a night attack. Now he simply sat on his horse and watched when Ishida's position came under strong attack, not committing his 1500 men. Ishida sent repeated messages to him, but still Yoshiro did not move a muscle. Finally Ishida himself came galloping up angrily and gave him a direct order to attack; Yoshiro blithely answered that he would do so when the time was ripe. Beside himself at this unlooked-for development, Ishida took some consolation from the brilliant showing of the regent Ukita Hideie who now started to make a breakthrough in his sector. In fact Ishida calculated that if he advanced much more, the point of the flank attack from Kobayakawa Hideaki might be lost. Yet already matters were taking on a momentum of their own. On Hideie's immediate left, Konishi Yukinaga's men were being pushed back by the easterners, and the battle bade fair to disintegrate into half a dozen separate conflicts. However, western hopes were buoyed by the strong showing of their troops on Hideie's immediate right.

Clearly it was time for Ishida to light the signal fires and order the *coup de grâce*. It was 11 a.m., three hours from the start of the battle, and the critical moment had arrived. Ishida lit the beacons to order Hideaki and his 15,000 down from the heights on to Ieyasu's flank. To the consternation of the western army, Hideaki ignored the signal. Even worse, so did Kikkawa Hiroie on the far flank; treachery was increasing with geometrical progression. When the impatient Mori sent scouts to ask Hiroie what was happening, he told them to go away as he was eating his breakfast and did not want to be disturbed. Unable to move from their positions without Hiroie's say-so lest they run into an ambush on the other side of the slopes, the Mori simply stayed put; what was supposed to be the key western strike force never even entered the battle. Meanwhile, Hideaki was openly revealing his treacherous hand. No matter how many envoys were sent up to him in his eyrie, he simply refused to move. Ieyasu noticed with relief that Hideaki's apostasy was now fact but began to worry when he did not move in any direction. Could it be that the young man was even more devious than he suspected? Was he hedging his bets, waiting to see the outcome before he made his move? Or was he even more chillingly ambitious, hoping that Ieyasu and Ishida would eat each other up so that he could emerge as the lucky third party and maybe even the new power in the land? Ieyasu did not know, but he was not going to let Hideaki get away with mere neutrality and refusal to commit. He ordered his harquebusiers to break off their assault on Hideie and Ishida and instead direct the full weight of their salvoes on to Mount Matsuo. The fusillade did the trick. Forced to go one way or the other, Hideaki gave the order to charge the extreme right wing of the western army.

IEYASU THE VICTOR

Almost immediately the entire battle was resolved. Heavily outnumbered and attacked from the front and flank, the brilliant eastern general Yoshitsugu was overwhelmed; in danger of capture he asked a retainer to run him through with a sword. Crashing past the scattered remnants of Yoshitsugu's men, Hideaki's turncoats next fell on the hard-pressed regent Hideie, utterly unprepared for such treachery. With enemies on all sides, rumours of treason and defection rife and cries of 'we are betrayed' rising on the air, the western front rapidly began to crumble. Vowing to take the traitor Hideaki's head before he died, the valiant Hideie tried to cut his way through to him, but his aides seized his bridle, restrained him and led him off the battlefield. By now only the west's left wing remained and the previously sullen and inert Yoshihiro finally saw mortal danger bearing down on him; he just managed to escape. On the face of it, Hideaki's defection, or rather Ieyasu's flushing him out with gunfire,

262

had turned the tide, but the really crucial thing in deciding the battle was Kikkawa Hiroie's inaction: his treacherous lethargy prevented all the people on Mount Nangu, the Mori clan and others, 25,000 in all, from taking any part in the conflict. This, even more than Hideaki's betrayal, was the key event. The unwilling non-combatants on Mount Nangu all scattered, and by 2 p.m. Ieyasu was completely victorious. With his generals in retreat and his own command post in danger of being overrun, Ishida accepted defeat and fled into the mountains.

Ieyasu pitched camp on the site of his great victory and spent the afternoon receiving the plaudits of his generals and viewing the heads of slain enemy leaders. The gruesome custom of viewing severed heads was an integral part of the Japanese military ethos, both confirming a victory beyond question and allowing the winning side to moralize about the defeated, either cursing and deriding a hated foe or praising the courage of those who had died valiantly. It is said that Ieyasu finally consented to don his ceremonial helmet now that the contest was decided. Scribes clustered around him, eager to hear the pearls of wisdom that dropped from his mouth or, more likely, put the words in there themselves. Certainly there is a slew of supposedly authentic post-Sekigahara sayings attributed to Ieyasu. 'After victory, tighten the cords of your helmet' was one such – in other words, don't be overconfident when you think you've won – but this seems to have been a traditional saw or apophthegm. More authentic-sounding are the ritual thanks expressed to his divisional commanders: 'Today's victory is entirely due to your loyalty and efforts. As long as my house flourishes, your clan will want for nothing.' The exchange with the Red Devils' leader shows the discourse of Japanese chivalry in action. A badly wounded Ii Naomasa was brought in, and Ieyasu insisted on binding up the wounds himself. Ii praised Ieyasu's son Tadayoshi, who had fought with him, but added: 'Hawks of fine stock always turn out well.' Ieyasu showed deftness in turning the graceful compliment: 'No matter how well-bred the hawk, a good trainer is always necessary.'

As the afternoon wore on, more and more petitioners for favour arrived at the camp. Kobayakawa Hideaki, who guessed that Ieyasu had seen right through him, prostrated himself at his feet and begged forgiveness, adding, in a face-saving gesture, that he regretted he had taken part in the siege of Fushimi castle under orders from Ishida. He was really asking that Ieyasu overlook his hesitation on the brow of Mount Matsuo that morning; the code was understood, and the pardon bestowed.

There was little magnanimity or forgiveness for those who had failed him, and still less for those who had opposed him. Much later in the afternoon, after all the heads had been counted and it was time to move on, Hidetada finally arrived. Ieyasu was initially minded to punish him severely and at first refused to see him or speak to him. Relenting later in the evening, he summoned his wayward son into his presence and reproached him bitterly: did he not realize that the loss of the 36,000 men he had diverted to chasing the will-o'-the-wisp of Ueda castle could have meant the difference between triumph and defeat? Hitetada was now formally forgiven but Ieyasu never forgot his son's idiocy and never entrusted important work to him again. A more difficult decision was what to do about the man-in-the-middle, Mori Terumoto: on the one hand he had taken no part in the battle (only because that was the way Ishida wanted it) but on the other he had agreed to be commander-in-chief of the west, an express act of hostility. Ieyasu despised Terumoto and made his contempt clear: it was simply not good enough to say that he was obeying Ishida's wishes, for Ishida could not even have gone to Sekigahara without the Mori clan; Terumoto's actions were classically dog in the manger. Yet Ieyasu did not want a renewed outbreak of warfare, with the unscathed Mori warriors still at large. He hit on a typically machiavellian way to punish Terumoto. Knowing that the turncoat Kikkawa Hiroie had done what he did to curry favour with the Mori, he simply confiscated Terumoto's most profitable fiefs and gave them to Hiroie, thus setting him at dagger's drawn with the Mori leader. Terumoto's fabulous wealth of 1.2 million koku shrank to a mere 360,000. When Hiroie tried to elucidate his motives and mentioned the assurances he had given Ieyasu before the battle, the victorious leader claimed not to know what he was talking about.

SEKIGAHARA: 40,000 SEVERED HEADS

Sekigahara was the greatest battle in the domestic history of Japan (it was eclipsed in overall Japanese history only by the two campaigns against the Mongols in the late thirteenth century, by the Russo-Japanese war of 1904–5 and the Pacific War of 1941–5) and ended the long period of civil war that disfigured the Nipponese islands in the second half of the sixteenth century. Maybe 50,000 men died on that fateful field, for Ieyasu's men claimed to have counted 40,000 severed enemy heads, and the east's own losses were not negligible. Yet although it was a great victory in terms of consequences, it cannot be counted among the great military triumphs of the ages, simply because Ieyasu's generalship had little to do with it; the battle was won and lost through treachery, pure and simple. He did, it is true, enjoy the advantages of greater discipline and a unified command and had timed the climax of his campaign perfectly, just before the winter snows would have put an end to campaigning. World history tells us that many of the greatest captains make their supreme effort in October for precisely this reason, and October is the cruellest month, since more great single battles have occurred in that month than any other. Without providing an exhaustive list, one might just mention the Greeks against the Persians at Salamis (480 BC), Alexander the Great at Gaugamela (331 BC), Octavian's defeat of

Mark Antony at Actium (31 BC), the battle of Hastings (1066), Henry V at Agincourt (1415), the Christians defeating the Turks at Lepanto (1571), the decisive sea battle of Quiberon (1759), the battles of Saratoga and Yorktown in the American War of Independence, Nelson's victory at Trafalgar (1805), Napoleon's at Jena (1806) and his defeat at Leipzig (1813), Balaclava in the Crimean War (1855), Caporetto in the First World War and El Alamein in the Second, and the sea battle of Leyte Gulf in 1944, to say nothing of the Cuban missile crisis in 1962 and the Yom Kippur war in 1973. If Ieyasu's timing was perfect, his decision to fight at Sekigahara can be faulted, for, without defections and treason, it is difficult to see how he could have avoided having the jaws of the western trap descend on him. On the other hand, given Japan's geography, he had few other realistic options at the time. Perhaps a more interesting question is, what was it at a deeper level that allowed Ieyasu to succeed? Here, apart from his undoubted political ability, one might cite the sheer length of his life (73 was a ripe old age to die in sixteenth-century Japan), his 11 capable sons, and the location of his power bases, both the original Mikawa-Owari region – a favourable location for dominating the imperial capital (all the three great unifiers had started from there) and the later Kanto 'grainbasket' district, with its agricultural wealth and productivity. Put simply, Ieyasu always enjoyed greater personal resources than his enemies, whom he could afford to despise for their factionalism.

NO MERCY

But though he could feel contempt, for his overt enemies there was no mercy at all. Ieyasu quickly advanced on Ishida's castle at Sawayama, which surrendered after two days. Ishida and his closest associate, Ankokoki Ekei, were both captured and executed. Legend has it that Ishida was kept in a cage until the day of his execution, but more securely based is the story that Ishida accepted his fate as karma and went to the execution ground at Kyoto with equanimity. When he and Ekei were being escorted to their beheading, a villager offered them some persimmons to eat. Ishida declined on the grounds of his digestion. When the others wryly pointed out that that was surely an academic consideration since he was going to die in an hour, Ishida quoted the Japanese equivalent of 'many a slip twixt cup and lip' and said that until the sword fell you could never know for certain what was going to happen. More interesting in the light of future developments was the fate of the rabidly Christian damyo Konishi Yukinaga. He turned himself in to his fellow Christian, Kuroda Nagasama, hoping he would get good treatment and be allowed to shrive himself and receive the last rites. But Nagasama feared that any concession to his coreligionist would simply bring down Ieyasu's wrath on himself, and refused any Christian privileges to his captive. Curiously, he offered Yukinaga the opportunity to commit seppuku, though he knew perfectly well that suicide was forbidden by Christian dogma. In Nagasama's eyes he was denying one Christian confession and extreme unction so that the entire Christian community in Japan would not suffer. Perhaps he already discerned the way Ieyasu's mind was working: Nagasaki meant the extreme west, the west was the enemy, and Nagasaki also meant the overweening power of the Jesuits, whom Ieyasu was determined to cut down to size. The luckiest survivor of Sekigahara on the defeated side was the regent Ukita Hideie. He went into hiding in the remote province of Satsuma, where his loyal clansmen ministered to him. By the time his hiding place was betrayed to Ieyasu, his post-battle rage had abated. He sentenced him to death for the form of the thing, but then exiled him instead: Hideie lived to the age of 84 and died in exile in 1655.

The post-Sekigahara dispensation

Ieyasu was now 58, and not expected to last much longer because of his notorious obesity. A rival of that other sixteenth-century despot, Henry VIII of England, in terms of bulk and girth, Ieyasu by this stage in his life resembled an immense sumo wrestler, enveloped in priceless silk robes that made him seem like a gaudy mobile tent. He liked to wear a blue satin robe embroidered with many silver stars and half-moons, which suggested a magus or sorcerer, except that he would always carry a sword at the waist to remind the unwary that he was a warrior. Long eyelashes and a wispy beard completed the impression of a wizard perhaps given to charlatanry, and it was said that at Ieyasu's court there was much tittering behind the ruler's back. Rumoured to be so fat that he could not tie the girdle that secured his voluminous robes, in later life he could not even mount his horse unaided. Cold and humourless, he was highly strung and operated on a short fuse. His habit of biting his nails when under stress has been much remarked on, and it was also noticed that as he directed a battle from horseback – he claimed, somewhat implausibly, to be a veteran of 90 battles – he would pound the pommel of his saddle until the blood ran from his hands; he was said to have developed a malformation of his fingers from this practice, with the middle joints calloused and rheumatic, so that in old age he had stiff and unbending hands. A believer in sumptuary laws and clothes that distinguished rank, Ieyasu also insisted on complete deference from underlings. At his court servants entered and left his presence on their hands and feet and acted in total silence with the utmost reverence; anyone not bowing low enough in his presence was liable to have his head lopped off by one swift cut from a samurai sword.

The immediate result of the post-Sekigahara dispensation was a massive expropriation and redistribution of the losers' property –

the greatest land transfer in Japanese history. Ieyasu took land worth 6.5 million koku from 90 of the vanquished and allotted it to his cronies; Hideaki was lavishly supported but died two years later, having barely entered his twenties. Hideyori (Hideyoshi's son) lost two-thirds of the domain left by his father but retained Osaka castle. Ieyasu was clearly in a position where he could have taken Osaka castle also but hesitated to attack the Toyotomi head-on; too many of his important allies had emotional ties with Hideyoshi and his family. Nevertheless, it was clear that the twilight area between the Tokugawa and the Toyotomi power factions and their general principles could not continue indefinitely. For a while Ieyasu decided to attempt peaceful co-existence. He gave Hideyori the provinces of Settsu, Kawachi and Izumi, with a revenue of 650,000 koku and in 1603 promised his six-year-old granddaughter to him in marriage. Ieyasu's generosity was doubtless a result of the other big event of 1603, when emperor Go-Yozei finally granted him the title of shogun — an accolade Hideyoshi had never been able to receive because of his low birth. The emperor also colluded in the fiction that Ieyasu was a descendant of the Minamoto family and granted him the corresponding honorific titles. Two years later (1605) Ieyasu officially retired as shogun and handed the title over to his son Hidetada. He called himself 'ogosho' (retired shogun) and withdrew to Sumpu castle, leaving Hidetada nominally in charge in Kyoto. The 'retirement' fooled nobody. Ieyasu was still pulling the strings but thought two layers of authority would confuse his hidden and latent enemies. By nominating Hidetada openly as his heir, he both established the precedent of direct succession and ensured that the claims of Hideyori's faction were nipped in the bud, preventing a legitimate succession struggle.

TAKING PLEASURE IN POWER

Ieyasu never relaxed his grip on Japan, playing elaborate games of divide and rule, fomenting the jealousy of one damyo against another, always watching for any sign of partiality towards Hideyori from the veterans of Sekigahara. After 1600 those who had pledged themselves to Ieyasu before Sekigahara were called the 'fudai' damyo, to distinguish them from the 'tozama' lords, who had made their submission only after the battle. Ieyasu liked to dangle bait before the tozama, hinting that they would lose their semi-tainted status if they performed extreme acts of loyalty for him. He never entirely trusted Kato Kiyomasa, the loyal Hideyoshi vassal whose defection to the east had so shocked Ishida, rightly viewing his actions as personal pique towards Ishida rather than ideological commitment to him. He raised his wealth to 520,000 koku but balanced this largesse by making Maeda Toshinaga the real beneficiary of the post-1600 settlement, with a total wealth of 1,250,000 koku. But when he finally suspected that in any clash between himself and Hideyori Kiyomasa would back the Toyotomi family, he had him poisoned (in 1611).

At Sumpu meanwhile Ieyasu established a court of all the talents. While receiving daily reports from Kyoto from his agents, he made a show of retreating into private concerns, indulging in his favourite pastimes of hawking, swimming and swordsmanship; he liked to swim around his castle moat for daily exercise. This too was one of the great periods of Ieyasu as exponent of bisexuality, when he was clearly in the great Japanese tradition of shudo or pederasty; it was to become a hallmark of the Tokugawa Shogunate that most of the great lords (eight out of ten of adult males, according to a recent study) took young boys as lovers. He had a favourite catamite called Ii Manchiyo and was criticized for spending too much time with his juvenile bedfellows; one eighteenth-century Japanese scholar said that Ieyasu was 'a sage who could not control his sex life'. All of this was in addition to his heterosexual activity, for Ieyasu had 19 wives and concubines by whom he had 11 sons and five daughters.

EXPULSION OF THE CATHOLICS

Ieyasu also ended nearly a century of Christian progress in Japan when he abruptly expelled all Catholic missionaries in 1614. Missionary penetration into Japan was headed by the famous Francis Xavier (later canonized), friend and associate of Ignatius Loyola, founder of the Jesuit order, and the so-called 'Apostle of the Indies'. Xavier initially made Christianity palatable by translating its doctrines into a quasi-Buddhist terminology and found powerful damyos to be patrons of the new creed, especially in the western island of Kyushu. The damyos thought Christianity would bring them the benefits of Portuguese trade with China, and to encourage this the Jesuits inveigled the Portuguese civil authorities into letting their 'Great Ship' dock in Japanese ports. By the 1560s Christian missionaries were making progress, and the Jesuits finally took advantage of Japan's endemic civil wars to gain a toehold at Nagasaki, where in 1580 the Omura clan sold the Society of Jesus the town and environs 'in perpetuity'. The Jesuits scarcely helped their cause long-term by displaying Christianity at its most intolerant: they destroyed Buddhist temples and Shinto shrines and made it clear that they intended to hang on to Nagasaki by force, if necessary. Their only chance of achieving this, of course, was while Japan remained fragmented. But the hegemony of Hideyoshi changed all that. The subjection of religious institutions to central control was the aim of all three of the great unifiers, but Nobunaga had never controlled Kyushu. Hideyoshi did, and so the days of the Jesuits were numbered. In 1587 he confiscated Nagasaki, sent in his own intendant to govern it, banned all missionaries and gave them 20 days to leave the country. What had particularly angered him was the destruction of Japanese shrines and the proven genius of the Jesuits for intrigue. Hitherto the attitude to the Christians had been decided entirely by the local damyos, but those days were now over.

Hideyoshi, however, decided not to enforce his anti-Christian decrees provided the Catholics remained discreet. Two things hardened his attitude. In 1593 Spanish Franciscans arrived in Japan and, in their rivalry with the Portuguese Jesuits, raised the Christian profile once more to the point where a ruler could not simply turn a blind eye. Even more seriously, in 1596 the great Spanish treasure ship, the Manila Galleon, was wrecked off the Japanese coast. The Manila Galleon was Spain's way of dominating the Pacific. Starting from Callao in Peru, the galleon headed across the Pacific to the Philippines, utilizing the ocean's favourable westbound currents, then turned north before crossing the north Pacific back to Acapulco. Delighted by such a treasure trove, Hideyoshi ordered the confiscation of the ship's cargo but the Spanish pilot, hoping to browbeat the ruler with tales of Spain's might, threatened him with retaliation. Gradually Hideyoshi heard the entire story of Spanish conquest in the Americas, of Cortés, Pizarro and the rest of the conquistadores. It seemed quite clear to Hideyoshi from this that the newly arrived Franciscans were indeed the advance guard of Spanish imperialism, for had he not been told that the missionaries were always sent ahead as a fifth column, to soften up native populations before the Tercios (Spanish troops) arrived?

Hideyoshi accordingly ordered a wholesale programme of persecution and executions; in 1597 the 'Twenty Six Saints' of Japan were crucified in Nagasaki. It was fortunate for the Christians that Hideyoshi died next year and that Ieyasu for a long time had other things on his mind. Ironically it was yet another blunder by the Portuguese that drew them to his attention. Learning that Christians had given large bribes to members of the Tokugawa household, Ieyasu revived Hideyoshi's decrees, denounced Christianity as an alien superstition hostile to Japan and expelled all missionaries. This was in 1614. Ieyasu was further angered that year at the siege of Osaka when he found so many Christians among his opponents, and remembered that Christians had fought in large numbers for Ishida at Sekigahara. Nagasaki's 25,000 population was by this time almost all Christian, and the Jesuits did not intend to abandon them without a struggle. They went underground, and by 1621, five years after Ieyasu's death, there were still 36 Catholic priests in the country. Hitetada, Ieyasu's son, hit back with further persecutions – in the 'Great Martyrdom' of Nagasaki in 1622, some 55 Christians were executed. But it took the third Tokugawa shogun, Iemitsu, who ruled from 1623–51, to finish off Christianity in Japan. By 1644 there were no missionaries left, and the Portuguese were gone too, expelled by a decree in 1639 that effectively ushered in two centuries of Japanese isolation from the outside world.

A SIREN VOICE AGAINST THE CATHOLICS

One of the siren voices urging Ieyasu to deal severely with the Catholics was that of his European favourite Will Adams. In one of those truth-is-stranger-than-fiction historical episodes, an English sailor born in Kent became a high-status samurai in Ieyasu's Japan. A native of Gillingham, William Adams saw early service in the Royal Navy, served with Sir Francis Drake, went on an expedition that searched for the North-West Passage and fought the Spanish Armada in 1588. Employed by a firm of Rotterdam merchants to emulate Magellan's and Drake's round-the-world voyages, but with a commercial objective, William sailed from Holland in 1599 with a fleet of five ships, with orders to seek commercial advantage in Peru and

the Molucca Islands as part of a circumnavigation of the globe. Continually battered by storms and high seas, Adams's fleet was nearly
destroyed; only his flagship, the Liefde, reached the coast of Japan, in
urgent need of food and repairs.

When he made landfall at Bungo (modern Usuki) in April 1600,
the Portuguese and the Jesuits told Ieyasu that Adams was a pirate
and called for him to be crucified in accordance with Hideyoshi's
anti-piracy laws. Thrown in jail, Adams was then closely questioned
about his motives and experiences by Ieyasu on three separate occasions before the ruler was satisfied with him. Among the Liefde's
cargo were 19 bronze cannons and 5000 cannonballs, which were
offloaded by the Japanese and allegedly used by Ieyasu at Sekigahara
though, strangely, the only first-hand accounts of cannon in the
battle deal with Ishida's use of it. Warming more and more to his foreign protégé, Ieyasu ordered Adams to build a Western-style ship.
When an 80-ton surveying vessel was delivered by Adams, Ieyasu
was so delighted that he ordered the construction of a 120-ton ship,
with which he hoped to make voyages to Mexico and Peru.

He further ordered that Adams must be allowed constant access to
him: 'Always I must come into his presence,' Adams reported. He
was not allowed to return to England or even (until 1613) to leave
Japan, but was gradually given more and more honours. Finally created 'hatamoto' (banner man) by Ieyasu – one of the shogun's most
valued retainers – and rewarded with land, serfs and revenues (a fief
worth 250 koku), Adams became a man with whom to be reckoned
in Japanese politics, feared and loathed most of all by the Jesuits and
the Portuguese, because he revealed the religious split in Europe
between Catholics and Protestants and enlisted Ieyasu's support for
the latter. The Jesuits reported: 'After he had learned the language, he
had access to Ieyasu and entered the palace at any time.' Though
Adams had a wife and daughter in England, Ieyasu decreed that Will
Adams the English sailor was no more and that the reborn Lord Pilot

('anjin-sama') should remarry and settle in Japan. He married Oyuki, daughter of a bureaucrat in charge of highways. At Sumpu castle Ieyasu assembled a cabinet of all the talents, a kind of brains trust, including the Buddhist priest Tenkai, Hayashi Razan, a Confucian scholar who became his chief legislator, Ina Tadatsugu, an expert on administration and government, Goto Mitsusugu, his financial and currency wizard, and Adams, his chief adviser on naval matters and foreign affairs. Other distinguished visitors to Sumpu included Miyamoto Musami, the greatest samurai and swordsman (*kensei*) in Japanese history, yet another who was supposed to have fought valiantly at Sekigahara (with Ukita Hideie, though there is no unimpeachable historical record of this).

BREACHING THE LAST BASTION: THE SIEGE OF OSAKA CASTLE

Japanese feudalism reached a high pitch under Ieyasu, with all the well-known trappings: feudal lords doing homage for their lands and swearing fealty, serfs rather than slaves working the land and a plethora of supposedly impregnable castles. Pride of place went to the great fortress at Osaka, where Hideyori and the Toyotomi resided. This was the showpiece of Japan, a mass of bulwarks, battlements, crenellations and embrasures set off by a massive drawbridge and moat. Inside the castle was vast parkland that resembled the grounds of the later palace of Versailles, with entire miniature landscapes and the trees and plants of all four seasons kept in coexistence by intensive horticulture. For a while the new shogun allowed the opulence of Osaka to outshine his more modest castle in the east.

But all the time during the halcyon years at Sumpu, Ieyasu was watching Hideyoshi's son and heir. The year 1611 (the year he poisoned Kiyomasa) was the turning point in relations between Ieyasu and Hideyori, who was now reaching maturity. Wishing to sound out the youth on his ambitions and thinking, he had an interview with him at Osaka castle and was alarmed at what he learnt. With his worst fears confirmed, he insisted that all the damyos swear a personal oath of allegiance to him as head of Japan's military estate. Still trying to avoid open warfare, he tried to weaken Osaka castle by draining the huge bullion store there to pay for monuments in Hideyori's honour. But, finding the Toyotomi faction as strong as ever, he finally cracked in 1614 and opted for hostilities as the only way to secure his son's future. He managed to pick a fight by a piece of transparent coat-trailing, claiming he had been deliberately insulted by the wording on a temple bell Hideyori had ordered cast. He then laid siege to Osaka castle, eagerly abetted by all the damyo who had opposed him at Sekigahara and lived to tell the tale.

Yet though Hideyori had not won over any of the great lords to his cause, Osaka castle was a formidable nut to crack. Defended by 90,000 samurai – both the landless kind dispossessed by Ieyasu's reforms and those who had been on the losing side at Sekigahara and had always wanted a return match, rightly claiming they had only lost through treachery – and surrounded by a double circuit of walls 3 kilometres (2 miles) long and 30 metres (100 feet) high, this was Japan's greatest fortress. The defenders, many of them Christians angered by Ieyasu's decree earlier that year and so regarding this as the last chance to stop the ogosho in his tracks, fought with desperation and valour and for a long time Ieyasu could make no impression on them, even though he poured 180,000 troops into the battle and used lavish supplies of gunpowder. Probably his artillery barrage finally proved decisive. Always a shrewd psychologist, Ieyasu guessed that Hideyori's mother, the lady Yodo, might be the weak link. Identifying her quarters in the palace, he had his gunners concentrate on that section of the castle; the distressed woman soon begged her son to start peace talks so that the shelling might cease.

Realizing that the Toyotomi were daily growing weaker – they had to fight continually whereas Ieyasu could always bring in fresh troops to fight in relays – Hideyori agreed to a truce. It came just in time for Ieyasu, who was losing face and political credibility as his armies failed to put a dent in the Osaka fastness. Ieyasu consented to a truce on condition some of the moats around the castle were filled in. The foolish Hideyori agreed, realizing too late that Ieyasu had actually identified the strong point in the defence and was using the armistice to remove the chief obstacle to the attackers. When he felt ready, in May 1615, Ieyasu resumed the attack. This time he was successful; the castle was taken and gutted and the defenders slaughtered; Hideyori and his mother committed seppuku. At last all the barriers to a Tokugawa Shogunate were removed and the memory of the Toyotomi was destroyed. Feeling his life's work accomplished, Ieyasu died the following year. So passed probably the greatest single figure in Japanese history.

CHAPTER 6

Napoleon
The master tactician and military genius

WHILE NO MILITARY operation can ever be fully understood without its political context, this is particularly the case with the siege of Toulon in 1793 – the occasion when the young Napoleon Bonaparte first made his mark on the world. The French Revolution of 1789 began as a moderate backlash against the corruption and deficiency of the Old Regime of the Bourbon kings but, as with all revolutions, the radical element soon took charge.

AN EYEWITNESS TO REVOLUTIONARY FRANCE

Historians usually identify the watershed in the Revolution as 10 August 1792 when the Parisian mob attacked the Tuileries Palace, slaughtered Louis XVI's Swiss Guards and demanded an end to the monarchy; interestingly, the 23-year-old Napoleon Bonaparte was an eyewitness. Prussia and Austria, as the vanguard of royalist Europe, took it upon themselves to end 'red revolution' and swore to sack Paris in revenge for the perceived outrage to Louis XVI. The Prussians invaded France, but were overconfident that they could deal with the revolutionary levies. Instead, the People's Army defeated them at Valmy on 20 September. This event created a sensation. The great German writer Johann Wolfgang von Goethe said to his fellow-countrymen: 'This is the beginning of a new era in history and you can claim to have witnessed it.'

Valmy became a famous symbol of French resistance. The most famous Frenchman of the twentieth century, Charles de Gaulle, later wrote: 'How can we understand Greece without Salamis, Rome without the legions, Christianity without the sword, Islam without the scimitar, our own Revolution without Valmy?' Revolutionary France, which now swung hard left under the Jacobin party, defied royalist Europe to do its worst and deliberately trailed its coat by executing Louis XVI in January 1793. This was too much for Britain, which had remained neutral hitherto; in February 1793 it expelled the French ambassador and was rewarded by a declaration of war from the revolutionaries. France now stood alone against the rest of Europe: its enemies included Britain, Prussia, Austria, Russia, Spain and the Italian principalities of Naples, Tuscany and Piedmont.

But the French revolutionaries faced an enemy within as well as a host of foes abroad. For nearly two years, from August 1792, the Jacobins had the upper hand in French internal politics. They aimed at the destruction of the bourgeoisie and the systematic use of terror. But the political right, representing the new middle class that had benefited from the sale of Church, royal and aristocratic properties, bided its time, waiting to seize power, which it finally managed in 1794. Although the Jacobins were triumphant in Paris, they had to engage in bitter civil war to hold down the rest of the country. In the northwest, the clerical and monarchist Vendée rebels posed a serious threat, and in the south the main cities were seized by 'counter-revolutionaries'. During the Jacobin supremacy the port of Toulon experienced some particularly violent months, with leftists in control of the inner city and conservatives having the upper hand in the outlying and country districts. The commandant of the city, the Marquis de la Flotte, tried to steer a middle course between the factions but for his pains was taken out and hanged by the Jacobins in September 1792. In reaction the conservatives tended to draw their inspiration from the old moderate Girondin party, expelled by the Jacobins from the central government (National Convention) in May 1793. The Girondins fanned the flames of revolt in the south, vowing there would be no one-party government along the Mediterranean coast; this was the so-called 'federalist backlash'. By July 1793 the conservatives were firmly in power in Toulon. This was largely because extreme Jacobin rhetoric about the abolition of property had no appeal there. The local working class, enjoying steady employment in the dockyards and naval arsenal, were something of an aristocracy of labour and wanted peace and prosperity, not the terror and 'permanent revolution' of the Jacobins.

Yet the revolutionary armies of the Jacobins were formidable and soon moved decisively against both the Vendée of the west and the rebels of the south. While they laid siege to the Girondin rebels in Lyons and Marseilles, the bourgeois government of Toulon panicked and surrendered the city to the British. As part of a general project of blockade of key French ports by the Royal Navy, a powerful fleet had been at anchor outside Toulon and Marseilles. On 21 August a revolutionary army under General Jean-François Carteaux took Aix-en-Provence and began to close in on Marseilles and Toulon. In alarm the Marseilles authorities asked the Royal Navy, under Admiral Lord Hood, for help. Hood told them he could do nothing for them at present – he was too short of men – but sent to his allies in the kingdoms of Piedmont and Naples for troops. It was too little, too late: on 25 August Carteaux broke through Marseilles's defences and captured the city, unleashing an orgy of massacre and atrocity by the Jacobin troops, angry at what they considered high treason (*la grande trahison*).

THE PRIZE OF TOULON

However, Hood was much more interested in a similar proposal from Toulon, as this gave him the chance to capture the French fleet there. Traditionally, the French navy stationed its warships in four ports: Toulon, Brest, Rochefort and L'Orient. In 1793 the second biggest concentration of vessels was at Toulon. There were 24 ships of the line and 22 frigates (as against 39 and 34 respectively at Brest). While Brest was the premier naval base, Toulon was a close second, far ahead of Rochefort and L'Orient (in third and fourth places respectively). Lacking specific instructions from the Admiralty, Hood consulted with his officers (who included the rising star Horatio Nelson). They decided to accept the offer of the Toulon rebels, but on Hood's terms: the British were to have total control of town, port and hinterland in return for supplying grain to a starving population and paying for all transactions in coin, not the inflated and despised revolutionary currency of *assignats*, the bank notes used by the Jacobins.

These were stiff terms, but with Carteaux's army at the gate, the Toulon bourgeoisie really had no choice. There remained the problem of the French fleet. The French admiral, the Comte de Trogoff de Terlessy, was willing to hand it over to Hood, but he had lost all authority and credibility through his virtually permanent absence on 'illness'. The effective commander was his second in command St Julien de Chambord, a Jacobin sympathizer, who refused to surrender. On 27 August Chambord ordered his men to battle stations but they refused. On the one hand, they did not relish the prospect of coming between two fires, from the rebels in Toulon itself and from Hood's fleet, but, on the other (and more importantly), Hood had suborned them by promising to make good all their accumulated back-pay in gold coin. Finding that an initial trickle of desertions was turning into a mutiny, Chambord and

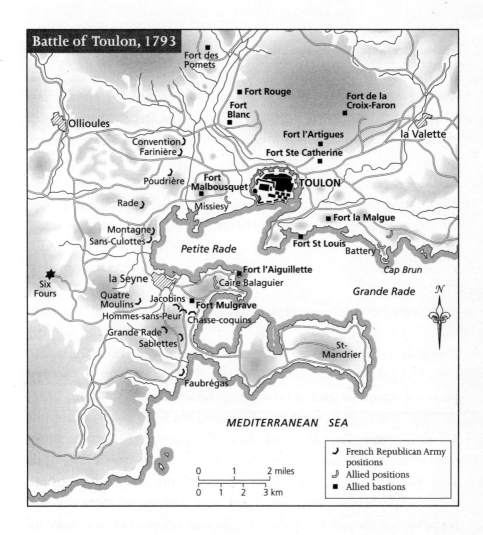

Battle of Toulon, 1793

his radical supporters threw in the towel and left overland to join Carteaux's army. Hood landed 12,000 British troops and 1500 sailors, plus about 1400 Spanish allies. They were only just in time. On 30 August Carteaux's army advanced to within 6.5 kilometres (4 miles) of Toulon, threatening that he would do to the city what he had done in Marseilles. But, once they had occupied all the key military installations and the heights around the town, the British were confident they could see off any Jacobin offensive.

They calculated that Carteaux did not have enough men to recapture Toulon, and meanwhile the rains and the chill of winter were coming on fast. British morale was further enhanced by an initial skirmish with the enemy, whom they repulsed easily. Long term, though, Hood made a number of mistakes, principally that of dividing his command between a governor of the town and a military leader, without making clear what the respective duties and responsibilities were. And Carteaux was quick to note that, even with 15,000 fighting men at his disposal, Hood lacked the manpower for a proper defence in depth around Toulon's huge and vulnerable perimeter.

A COMPLEX SETTING FOR A SIEGE

The geography of Toulon is complex, but the siege of 1793 would sound like gibberish without an in-depth knowledge of the topography of port, city and environs. Anyone approaching Toulon from the east by sea would round the headland at Cap Brun and find himself in the outer bay or Grande Rade. Sailing into the Grande Rade, one would immediately notice the waterway narrowing as the Saint-Mandrier peninsula loomed up on the west; this was steep and rocky, 120 metres (400 feet) high and with heavy batteries that commanded the whole of the Grande Rade. Soon a promontory on the eastern side ran out almost to meet Saint-Mandrier, funnelling any traffic into the narrow entrance to the inner bay or Petite Rade. Here too was another battery (Fort La Malgue), so that the two fortified promontories on either side of the entrance to the inner bay had the entire Petite Rade within range; any seaborne assault would therefore be suicidal. The outer bay was sheltered from the south and southwest and at its back were mountains; though no more than 600 metres (2000 feet) high, they were rugged and precipitous, with a

discontinuous ridge broken by gorges. The town and arsenal of
Toulon, on the eastern side of the Petite Rade, were thus in a sheltered position facing the bay and with mountains at their backs. A
defensive city wall ran round both town and arsenal, lapped by a
moat that was partially tidal. Within the port were two basins, the
old and new harbours. Adjacent to the former was the old town, a
maze of narrow streets where the dockworkers lived; next to the
new harbour was the middle-class residential area of St Roch. On flat
ground to the west of Toulon, 1.5 kilometres (1 mile) distant, was
Fort Malbousquet, covered by a battery about 1 kilometre (about
half a mile) to the rear on the rise at Missiesy. The Faron ridge to the
north of Toulon was particularly problematical for both attacker and
defender. Where Fort Faron was overlooked by the eastern end of
Mount Faron ridge, some 150 metres (500 feet) higher, Hood built
a fastness named Fort de La Croix Faron, at an altitude of 565 metres
(1850 feet).

A valley separated the western end of Faron ridge from its
neighbouring height, Le Croupatier; though wide, the valley was
constricted by a series of spurs at its northern end – and these spurs
were commanded by a minor road that ran through the village of Les
Moulins. Some 1.5 kilometres (1 mile) away, Fort Pomets crowned
the southernmost spur but was not well placed to cover the routes
from the north, so the later Redoute Saint-André had been added to
cover both Pomets and the nearby road junction. As this route was
an obvious weak point, further strongholds had been added, whose
names recur with monotonous frequency in the copious revolutionary dispatches: Saint-Antoine, Petite Sainte-Antoine, Fort Rouge and
Fort Blanc. So much for the eastern and northern side of Toulon
bay. But it was the southern and western side that was to prove the
most vital sector in the coming conflict. Following the road round
from Fort Malbousquet to the village of La Seyne and beyond, one
came to the western promontory that covered the Petite Rade, about

3 kilometres (2 miles) across the bay from Toulon town. This was the peninsula of La Cairne, and here stood the forts of L'Aiguillette and Balaguier, the most crucial of all links in the defence, as they commanded both the town and the Grande and Petite Rades. Eventually Hood came to see the importance of these strongholds and ordered work to start on a further defensive screen to their rear that came to be known as Fort Mulgrave, but only after being alerted by a French attack. It is fair to say that 'Fort Mulgrave' is a creation dating fairly precisely from 22 September. Ranged around a 90-metre (300-foot) eminence called Hauteur de la Grasse or La Cairne, this complex set of positions finally seemed so impregnable that the French under Carteaux called it 'le petit Gibraltar'. From Fort Mulgrave the route to the further battery at Saint-Mandrier peninsula ran along the narrow isthmus of Les Sablettes.

STALEMATE AT TOULON

Hood's dispositions for the defence of Toulon were admirable, and he was particularly good at guarding against a possible approach from the east (from the French Revolutionary Army of Italy). Around Fort La Malgue and to the east of the town walls he occupied the positions of Saint-Louis, Grosse Tour and Sainte-Catherine and also installed a garrison at L'Artigues to the northeast (in what is now the Corniche des Farons). If the eastern chain of communications were solid and formidable, the position on the western side of the bay was less secure. Fort Malbousquet, positioned to prevent any attack from Ollioules, was isolated, too far from Fort Mulgrave and thus vulnerable to being outflanked. Carteaux was no great general but even he could spot that the weakness of the British position was in the complex of La Cairne and the two batteries at Balaguier and L'Aiguillette. If he could take them, or even if he could find a way to get down on to the Saint-Mandrier peninsula, he would make Hood's position in Toulon untenable.

The question immediately arises: why did not the British also spot the obvious danger? The answer is that they did but were forced to neglect the western sector for two reasons. In the first place, Carteaux did not initially have the numbers for any meaningful assault on Toulon, but it was likely he would ultimately be reinforced from the Army of Italy, which meant a threat on the eastern side. Hood also considered an amphibious assault from the west unlikely, for this would mean that the Jacobins had an unreliable Marseilles in their rear, which might well rise up against its oppressor in such a situation. Yet, most of all, Hood was constrained by manpower shortages. Already his troops were stretched thin around the defensive perimeter, and the issue of reinforcements was tricky. Prime Minister William Pitt had already cut the strength of the British army and most of its regiments were earmarked for service in the West Indies, which meant that any fresh troops would have to come from the allies or be bought in as mercenaries. The underlying problem was that Toulon was never a priority in London. Hood should really have removed the French fleet from Toulon to Minorca or Gibraltar, but he dithered, fearing that such decisive action would turn the people of Toulon against him.

The obvious French strategy was to outflank Fort Malbousquet, occupy La Seyne and then mass for an assault on Balaguier and L'Aiguillette; if they seized the western shore of Toulon's inner harbour, the town and the British would be done for. On 7 September, Carteaux made his move. Probing carefully down the Gorges d'Ollioules, he committed his hardened veterans to the inevitable fight for the defile at the southern end of the gorge. He ran into stiffer opposition than expected. First, the Toulon National Guard made a spirited showing and held off their revolutionary brethren for five hours, taking two-thirds of their number (about 70) as casualties. Then, Spanish infantry came to the rescue but were also severely mauled by Carteaux's men, again losing about 50 in casualties. Eventually, Carteaux was forced to retreat but he had had the better of the argument and forced the allies to garrison the Ollioules front more strongly, inevitably weakening other sectors.

Yet he had come nowhere near solving the problem of how to take La Cairne peninsula. Correctly reading his strategy, Hood ordered his naval vessels to position themselves for maximum support for Balaguier and L'Aiguillette. Realizing that his larger warships could not operate in the shallow water at the western end of the Petite Rade, around the shoreline of La Seyne, he decided to deploy lighter craft there. He ordered the construction of two large pontoons, each armed with four long 24-pounders and two brass mortars; additionally, he strengthened La Cairne by landing big guns taken from the immobilized French warship *Aurore*. By the second week of September the siege of Toulon was at stalemate, with Hood worrying about manpower and Carteaux doing no more than firing desultorily on Malbousquet.

A CORSICAN PATRIOT IN TOULON

This was the precise moment when Napoleon came on the scene. Born in Ajaccio, on the west coast of Corsica, in 1769, the second son of an impoverished lawyer, Carlo Bonaparte, Napoleon had been educated at the military schools of Brienne and Paris, and in 1785 was commissioned as a second lieutenant of infantry in the regiment of La Fère, garrisoned at Valence. Always politically ambivalent, prepared to run with the hare and with the hounds, Napoleon started as a Corsican patriot and even endangered his career by returning to the regiment late in order to dabble in the politics of his island homeland. But he fell foul of Pasquale Paoli, another political trimmer posing as a patriot, but then masquerading as the father of Corsican liberty. The entire surviving Bonaparte family (Carlo was dead by now) had to decamp with speed from Corsica and arrived in Toulon in the summer of 1793 (before the surrender to the British) to find the Terror at its height.

Virtually penniless, the Bonaparte family was saved by the intervention of Napoleon's old friend Christophe Antoine Saliceti, the leader of the Corsican faction Napoleon had, unsuccessfully, backed against Paoli. Saliceti had powerful connections in the National Convention in Paris and put in a plea that the Bonapartes, with all their property in Corsica expropriated by the Paolistas, had suffered grievously for the Revolution. The Bonapartes were voted a money grant of 600,000 francs, which seems never to have been paid, but the family fortunes recovered now that Napoleon and his siblings were tagged as Jacobin heroes (his brother Lucien was in fact the only true leftist). Moreover, Napoleon's talent as a pamphleteer and master of propaganda had brought him to the attention of Augustin Robespierre, brother of the more famous leader of the new 12-man government in Paris. The alliance of Augustin Robespierre and Saliceti did wonders for the 24-year-old Napoleon, for he now had friends and backers in the highest circles of government.

Saliceti saw a chance to catapult his protégé into early promotion. As a lieutenant of artillery, Napoleon was ordered to escort powder wagons from Marseilles to Nice, ready for use by the Army of Italy. Saliceti then set it up that Napoleon should stop at Beausset to pay his respects to him and his colleague Gasparin. Saliceti and Gasparin were both *députés en mission* – political commissars with wide-ranging powers to oversee army generals who might have suspect political loyalties. It so happened that General Carteaux had recently lost his artillery commander Dommartin, invalided out with serious wounds. Saliceti suggested to Carteaux that he employ Napoleon as the replacement. Carteaux was reluctant, but dared not oppose the will of a political commissar, lest he be denounced as an 'enemy of the people'. Carteaux was right to be suspicious of the newcomer, for Napoleon took against him on sight and regarded him as a military incompetent. He had 12,000 troops under his command and another 5000 detached from the Army of Italy under General Lapoype but had no real clue how to take Toulon. Carteaux was one of a long list of people who detested the young Corsican – a list that would lengthen considerably over the next 20 years.

Napoleon was a classic love-him-or-hate-him personality, at once attracting fanatical devotion and loyalty from his supporters and visceral loathing from his enemies. The loathing even extended into his family, for though three of Napoleon's brothers admired him and were in awe of him, the fourth, Lucien, was both insanely jealous and deeply resentful of him. Immediately sensing that Carteaux reciprocated his low opinion, Napoleon at once set out to discredit him with the government in Paris, giving Saliceti and Gasparin (he too was a devoted Napoleon admirer) snippets of damning information, which they duly passed on to the Robespierres in Paris.

EARLY SIGNS OF MILITARY GENIUS

Napoleon began by going down to the French lines near La Seyne to reconnoitre. Although isolated individuals on both sides had from time to time suggested that the La Cairne peninsula was the key to operations, until the Corsican's arrival no one had seen with total clarity just how vital it was. Hood and the British had actually disarmed the old French guns there and abandoned the position at the end of August, thinking their warships were sufficient to counteract any threat in this theatre. Until mid-September they were right. Carteaux concentrated on investing Toulon on all sides, closing the ring by shrinking the perimeter around town and bay. The problem was that trying to reduce Toulon by such methods would have resulted in a siege as long as that of Troy, even supposing the National Convention were prepared to commit the 150,000 troops Carteaux thought he needed for the tasks. Carteaux's plans were vacuous as they both ignored the peculiar geography of Toulon – meaning that at all costs concentration of force should be brought to bear on L'Aiguillette – and presupposed manpower levels that were simply not available.

Napoleon saw at once that the fall of the La Cairne peninsula would have a multiplier negative effect on British morale: on the one hand, they liked to fight with their backs to the sea, and on the other, they would be aware that with the fall of L'Aiguillette the French could interdict all seaborne assistance, making the fall of Toulon inevitable. The young artillery lieutenant's original insight was that only artillery could break down these particular walls of Jericho, and that Carteaux should accept the principle that artillery should form the spearhead of the attack, with infantry in support, rather than, as in classical tactical manuals, artillery playing an ancillary role to the infantry. If artillery were to be the breakthrough, its power would have to be overwhelming if it was to best both the defences of 'little Gibraltar' and the firepower of the Royal Navy warships. Napoleon therefore argued that the full resources of the military should be spent on artillery batteries.

On 18 September he opened fire from a masked battery that he had erected from behind a screen of pine trees to the northwest of the inner road. He invited the representatives to view his work, and Saliceti duly admired it; Fréron was absent in Nice. The artillery bombardment of La Cairne seems to have come as a complete surprise to Admiral Hood, but he reacted quickly, employing a number of floating batteries with heavy cannon, gunboats and low-draught rowed galleys to flush out Bonaparte's batteries. This artillery duel continued until 19 September, with the French at first having the better of the action. But by the evening of the 19th the republican batteries were silenced, albeit for the loss of a couple of British gunboats. Mulgrave at last saw that this sector was the key to the entire campaign. 'The road would not be tenable for the fleet if the enemy should take possession ... it became absolutely necessary to collect a force to occupy it.' Mulgrave in turn took swift action and landed 500 Spanish and British troops at Balaguier on the night of 20–21 September. These soldiers laid the foundations of what would later become Fort Mulgrave by cutting down trees and forming a rough breastwork in front of them. At dusk the French attacked with 400 men, unsupported by cannon or cavalry; they were easily repulsed. So feeble was the attack that Carteaux was later accused of deliberately failing, to 'prove' that Napoleon's ideas for a *coup de main* were impracticable (he hoped that no one would look too closely at the calendar). At dawn on the 22nd the Allies began the formal construction of what would become Fort Mulgrave or 'le petit Gibraltar'. From this moment on, the La Cairne peninsula was no longer vulnerable to a *coup de main*.

It is worth stressing that although desultory attempts had been made to land men on the peninsula before mid-September, only Napoleon had seen that with artillery it was possible to take the position quickly. Carteaux later claimed that Napoleon's contribution to the siege strategy was nothing special, as he had himself seen the importance of what became Fort Mulgrave all along. In which case, the obvious question arises: why did he not simply march his men down there and occupy it? Hood and Mulgrave, too, realized what an egregious error they had made and tried to fudge the record by claiming that the landing of Spanish troops was a reinforcement, not an initial occupation. By his brilliance in grasping both the importance of the peninsula and the most effective way to take it (by concentrated artillery fire), Napoleon had unwittingly given the game away. For the rest of his time at Toulon he continued to fume about the lost opportunity. On 14 November he was to write to the Minister of War in Paris: 'Had it [his original plan] been followed from the beginning with a little more ardour, it is probable that we should now be in Toulon.'

Saliceti wrote earlier, on 25 September, impugning Carteaux:

The peninsula remained totally ungarrisoned until Bonaparte, denied permission for his coup de main by the sceptical Carteaux, by erecting batteries to drive the allied fleet from positions near the peninsula, alerted the allies to the value of the position. In order to occupy and hold an empty peninsula it seems reasonable to require only 'a handful of men', and reasonable for a military genius to grasp this.

Napoleon was furious that the enemy had reacted so quickly and were now building a sturdy fort. 'They are going to receive considerable succours and we must make our minds up for a siege.'

AN ARTILLERYMAN OF ORIGINALITY AND GENIUS

Naturally, all of Bonaparte's thinking, whether about *coups de main* or artillery, was heresy to Carteaux, who resisted, but Napoleon was once again able to overawe him with the help of Saliceti and Gasparin. His problem was that he badly needed more resources, as at present the French military arm consisted merely of eight 24-pounders, three 16-pounders and two mortars. He had long been an apostle for the doctrine of massed artillery. When the La Fère regiment was stationed at Auxonne in 1788 he had devoured the handbook written by General Jean de Beaumont du Teil that stressed the massing of big guns at decisive moments in battle. Napoleon believed that the concentration of firepower at a single, vital sector could offset even severe numerical superiority in infantry. The next weeks saw frenzied activity as the bases of an armoured division were assembled. Napoleon, a dynamo of energy, assembled 100 pieces of artillery, mostly medium-calibre 24-pounder long cannons and heavy mortars. At Ollioules he assembled horses to drag the big guns and constructed his own arsenal, with 40 artisans – blacksmiths, wheelwrights, carpenters – working night and day on what he needed. He built a foundry and a gunsmith's workshop where muskets could be repaired. Notable was his habit of scouring Provence far and wide for what he wanted. From Marseilles he enlisted a gang of men weaving baskets and wickerwork for demijohns and making gabions; additionally 5000 earth sacks a day were made in Marseilles. He requisitioned horses from as afar afield as Nice, Valence and Montpellier and used up all the wood in the environs of La Seyne and La Ciotat.

Napoleon had one great advantage. By 1793 France was ahead of the rest of Europe in the technology of artillery. Bruised after the decisive defeat by Britain in the Seven Years War of 1756–63, France had thereafter begun a massive programme of upgrading its technical abilities in warfare. As a result of the innovations of the engineer Jean-Baptiste de Gribeauval, field guns had lighter barrels and carriages, making it possible to produce 12- or 24-pounder calibres, a level of ordnance hitherto thought feasible only with heavy siege guns. The famous victory at Valmy in 1792 was the most advanced battle so far in terms of big guns and artillery rounds fired. The war fever of 1793 saw massive production of artillery weapons – 7000 cannons in that year alone – and the efforts of scientists like Gaspard Monge ensured that France maintained its technological lead. All that was needed for France to have clear superiority in artillery duels was an artilleryman of originality and genius – virtually a description of Napoleon Bonaparte in 1793. Napoleon grasped that in a close-range contest of ordnance he would have the edge, provided he could keep enemy musketry at bay. Here he had the advantage that musketry was only really accurate at less than 45 metres (50 yards) range. Napoleon therefore issued his attack squad with rifled carbines – lighter, smaller-calibred weapons – which enabled snipers and skirmishers to take a heavy toll of enemy infantry while still outside effective musketry range. While he unleashed a devastating bombardment from his big guns, he would send dense clouds of skirmishers forward to pick off key enemy officers and sow chaos and confusion. As a rule of thumb, Napoleon liked to have four big guns for every 1000 infantry.

'LYONS MADE WAR ON FREEDOM. LYONS NO LONGER EXISTS'

While Napoleon breathed new life into the hitherto lacklustre campaign of General Carteaux, in Toulon Admiral Hood was running into more and more problems. From the Admiralty came an order replacing his original choice as governor of Toulon with General Charles O'Hara, then based at Gibraltar. From his spies came worrying reports of the 5000 captured French seamen, all staunch revolutionaries, who were a potential fifth column and whom he could not afford to police properly. Finally exerting himself, Hood detached four of his least serviceable warships, loaded on the French seamen and (on 17 September) shipped them out for repatriation at Rochefort and Brest. Maybe this was machiavellianism, for four days before the sailors in the Brest fleet had mutinied, causing the National Convention in Paris another headache. Yet slowly but surely events were turning in the Jacobins' favour.

The revolutionary armies were slowly closing the ring around Toulon, with General Lapoype and the Army of Italy moving in from the north and east. The political commissars Barras and Fréron, who were accompanying the Army of Italy forced the citizens of

Marseilles to disgorge a 4 million livres 'loan' (after the recent atrocities the Marseilles people scarcely dared refuse), so that Carteaux and Lapoype – and in consequence Napoleon – had all the money they needed. Mobilization orders went out to every able-bodied male between the ages of 16 and 60 in the Département de Var, and Napoleon was able to press into service all recently retired artillerymen in Provence. The Toulon campaign was becoming part grudge match and part test of credibility; the Jacobins knew that if they did not crush Toulon quickly, resistance would increase elsewhere and the flames of the Vendée would be fanned into a general nationwide civil war. Morale in revolutionary quarters rose dramatically at news of the fall of the royalist city of Lyons in early October. Once again the most horrifying reprisals were visited on the defenders. The victorious Jacobins set up a column bearing an inscription that read: 'Lyons made war on freedom. Lyons no longer exists'.

Meanwhile, Napoleon had completed his first two battery sites near Bregaillon, a hamlet just north of La Seyne, with clear arcs down the Petite Rade. Following the revolutionary habit of giving 'inspirational' names to everything, he called his two batteries the Sans Culotte and La Montagne, names with a clear revolutionary ring; soon they were joined by a third and all three just had the range to hit Fort Malbousquet. His prize gun was a 44-pounder culverin, with a barrel 25–30 times its bore (as against 18 normally). An expert on gunnery, Napoleon knew that the extra length of this weapon ensured greater accuracy, range and consistency of aim. His first target was Hood's pontoons, and a fusillade on the night of 17–18 September alerted the British to the presence of some formidable French weapons; unfortunately, they lacked the manpower to land a raiding party to take out the guns. Hood ordered the 98-gun *Saint-George* to come as close to the shore as it could and unleash a broadside, but it is a proven axiom of warfare that naval gunfire can rarely dominate over land-based artillery. Gunners in the Royal Navy were trained to fire rapid broadsides at short range, not the accurate, precision long-range bombardments in which Napoleon specialized, and this was quite apart from the elementary consideration that ships, even at anchor in the bay, rolled and pitched whereas their shore-bound adversaries could deploy steady, unrocked fire.

With Napoleon's gunners using red-hot shot (the so-called *boulets rouges*), the overall military grudge match was counterpointed at the artillery level with an exchange of cannonades that went on virtually non-stop from mid-September. Napoleon's next step was to tighten the noose on Fort Malbousquet. French probes netted a number of small hills, and on these he sited three more batteries: the Redoute de la Convention on the eminence of Les Arènes, the Farinière on La Goubau heights and the Poudrière on the Hautes des Gaux. Bit by bit Napoleon began to establish a mastery on the western shore of the Petite Rade. It was no wonder that on 30 September Gasparin and Saliceti reported to their political superiors in Paris that Napoleon 'is the only artillery captain able to grasp operations'.

TOULON IN THE BALANCE

Admiral Hood grew increasingly concerned that the allied garrisons on the western promontory and the Saint-Mandrier peninsula were becoming isolated; now that Napoleon's gunners commanded the no man's land between Fort Malbousquet and Fort Mulgrave, the western redoubts were dependent for reliable communication on the Royal Navy's command of the water. It was quite clear that the French tactics were to keep the British and their allies stretched on the eastern side while they pounded the positions on the west, waging a grim, slugging battle of attrition by gunnery. Hood desperately needed more manpower, but now the deficiencies of the grand strategy of Pitt and Henry Dundas at the Board of Control became clear.

Their policy of opportunism, creating numerous 'hotspots', had been too successful; they had a number of valuable strategic plums in their hands but lacked the manpower to hang on to them. Hood was initially promised 5000 British reinforcements to hold the line at Toulon, but after opposition from the Duke of York this order was rescinded. Hood's only option now was to beg more troops from Britain's allies in the Mediterranean, but he was deeply unhappy with their quality: he referred to the Neapolitan soldiers as 'trash'. The obvious source for reinforcements was Spain, but Madrid had entered the war half-heartedly against its 'natural' ally France and remained lukewarm and suspicious of Britain's intentions in the Mediterranean. Thinking the calibre of Austrian fighting men superior, Hood tried to get troops from that quarter, but Vienna refused. Foreign Secretary William Grenville, Pitt's cousin, did promise (in mid-September) to transfer 5000 Hessians (famous or notorious from their activities in the American War of Independence) from Flanders, but this pledge too was never honoured.

With the Spanish admiral don Federico de Gravina a reluctant co-worker with Hood, the prospects for a long-term retention of Toulon by the allies seemed dim, but finally Hood was categorically assured that reinforcements were on their way from Gibraltar and that Austria had finally agreed to send their 5000 men. Hood's field commanders tended towards complacency, lulled by Carteaux's inactivity and the general atmosphere of 'phoney war' prevailing, apart from the non-stop artillery duel with Napoleon's batteries. More solid support finally seemed assured when Sardinia agreed to send 20,000 troops to Toulon in return for a £5-million subsidy and help in the recovery of Nice and Savoy; even though this agreement was formally ratified, in the event only 2000 Sardinians arrived, whereas the dispatch of the full complement would have made Hood's position unassailable. With all his manpower shortages, Hood foolishly agreed to a request for help from Napoleon's old enemy Paoli in Corsica and his 'patriots'. Hood sent three warships under Commodore Linzee with three warships and two frigates, but he accomplished nothing and fared badly in Corsica, being let down by the unreliable Paoli; ironically Hood came to have the same opinion as Napoleon of the 'father of Corsica'. Another irritant for the hard-pressed Hood was the behaviour of Genoa. Officially neutral, the Genoese were guilty of several severe infractions of their alleged status, so Hood retaliated in like manner, boarding French warships in Genoese waters. When Genoa threw off the mask of neutrality and declared for France, Hood ordered the port blockaded, but the Austrians used this as an excuse not to send the promised 5000 men, on the slender pretext that Genoa was to have been the transit port.

THE 'CORSICAN UPSTART' AND FRENCH FACTIONALISM

While he laboured daily to build his artillery up to the point where its combined firepower would be devastating and irresistible, Napoleon was increasingly frustrated by the inactive and lethargic Carteaux. Napoleon, backed by Saliceti and Gasparin, deluged the National Convention in Paris with complaints about their general, but Carteaux on paper had a string of victories to his credit and the Convention was reluctant to act, even at the urging of their political commissars. Relations between the general and his director of artillery reached breaking point as the angry and jealous Carteaux made his resentment of the 'Corsican upstart' ever more obvious. But Carteaux scarcely helped his cause by falling out also with Lapoype, commanding the Army of Italy. The *députés en mission* accompanying Lapopype, Paul Barras and Fréron, were also pro-Bonaparte, so the luckless Carteaux now had four political commissars telling the Robespierre brothers how useless he was. Both Barras and Fréron would play major parts in Napoleon's subsequent history:

Barras as a senior figure in the later governing body, the Directory, and Fréron as the lover of Napoleon's delectable and lubricious sister Pauline.

Angry at the whispering campaign being conducted against him by the commissars, Carteaux decided in effect to work to rule. When Hood, deeply worried about the increasing strength of Napoleon's batteries, landed a force of 600 British and Spanish troops from Toulon to shore up the western promontory, Carteaux deliberately sent an under-strength force against him, which was predictably defeated. This then enabled Carteaux to write to Paris that 'Bonaparte's plans' had miscarried. Elated by the defeat of Carteaux, Hood on 23 September ordered the combined fleets to blast the French artillery batteries off the face of the western shores. Some 11 hours of spectacular bombardment followed, with one Spanish warship alone using 1700 rounds. The allies reported the batteries silenced, but Napoleon thumbed his nose at them by opening up a fresh fusillade next day with his only slightly damaged guns.

The end of September saw both sides enjoying mixed fortunes. Some 2000 Neapolitan troops finally arrived to give Hood's defenders a shot in the arm, but the French were meanwhile riven with factionalism. On the eastern side Carteaux joined hands with Lapoype's army beyond the Mount Faron ridge, thus tightening the cordon around Toulon, but immediately the two commanders once more fell out over strategy. Carteaux wanted to take Cap Brun, on the easternmost extremity behind La Malgue, looking directly across to the Saint-Mandrier peninsula, which would further tighten the net and cut down on allied anchorage in the Grande Rade. But Lapoype, backed by all the commissars, decided the next move would be made against the Faron heights; it can thus be seen that both Carteaux and Lapoype were ploughing their own furrows and deliberately cutting down on the assistance available to Napoleon's artillery – the only real chance of success. Although Fort de la Croix Faron at the very highest point of the ridge could only be approached from the eastern side by sheer cliffs, Lapoype's alpinists performed most valiantly and, after climbing up vertiginous goat tracks, swarmed over the totally surprised stronghold.

When Fort Faron, now overlooked by the invaders, signalled for help on 1 October, the British set off with 1200 troops to try to clear the French off the heights. After struggling to the top, the weary British faced a French force perhaps 1800 strong in the vicinity of Fort de la Croix Faron. But the British were crack troops and the defenders a scratch force. There followed a decisive allied victory, with the French losing 700 dead – shot, bayoneted or perishing by falling off cliffs. Only bad light prevented a genuine massacre, and this signal triumph was achieved for the cost of just 11 allied dead and 70 wounded. Predictably, Carteaux did not miss this opportunity to point out to the Committee of Public Safety in Paris that his ideas had been right and Lapoype's disastrously wrong, but Lapoype still enjoyed government backing and lived to fight another day.

CHAOS IN THE COALITION RANKS

With 2000 Neapolitans having newly arrived and the Sardinians and Piedmontese performing well, the British might now have been euphoric about their prospects had it not been for the Spanish. Grossly indisciplined and inclined to turn to outright brigandage if they did not get what they wanted, the Spanish troops severely alienated Hood and his officers, but national pride made Gravina and the other Spanish commanders close ranks in support of their own, with a consequent plummeting in relations between Spain and Britain. By the first week in October even some of the Spanish officers thought the chaos in their own ranks had gone too far, and one major wrote about the senior commanders: 'One does not know who commands, for each pulls in his own direction.' It was becoming clear that the coalition was at best a heady brew – and how not, given that English, Neapolitans, Piedmontese, Spanish and even Swiss mercenaries had their own agenda.

But Hood's best general, Lord Mulgrave, who had commanded at the victory of Fort de la Croix Faron, was so buoyed by his troops' showing that he decided to strike straight at Napoleon's batteries, particularly targeting one south of La Seyne village, one at Quatre Moulins and two on elevated ground at Regnier. On the night of 8–9 October Mulgrave led a sortie in force from Fort Malbousquet, which made some initial headway and spiked a few guns before falling back again. Five days later the French, elated by the news of the fall of the royalist stronghold at Lyons, counter-attacked and fought a brisk skirmish near Malbousquet. They failed to take the fort but demoralized the defenders by spreading word that those who were no longer needed at Lyons would be heading south to join Carteaux's army.

Carteaux should have kept up the pressure in this sector but allowed himself to be diverted when Lapoype sent word that he was now interested in the very assault on Cap Brun that Carteaux had previously advocated, but which had been overruled in favour of the attack on the Faron heights. On 15 October, 2000 French troops surprised the garrison at Cap Brun but had made no plans to hold the position against a counter-attack. Mulgrave reacted strenuously. Quickly perceiving that the French had stretched their lines of communication almost to snapping point and would be vulnerable to a counter-attack on a line of retreat through La Garde and Thouar, he tried to catch them in a pincer movement, sending one column to retake Cap Brun and another to La Vallette on the Thouar road to intercept the retreat.

The pointlessness of the French capture of Cap Brun soon became evident. Hearing that Mulgrave was marching on them from Fort La

Malgue, they decamped at speed back along the road on which they had come and thus passed through La Vallette before the Spanish troops could cut the line of retreat. Mulgrave very sensibly sited a battery on the heights of Thouar to make sure the same ploy could never be tried again, but spoiled his efficient actions by two political blunders. In the first place, he complained to Hood, unfairly, that only Spanish lethargy had allowed the French to escape his trap; inevitably, news of this complaint leaked out, further exacerbating Anglo-Spanish relations. Even worse, he sent dispatches to London, boasting of his prowess, whereupon the Pitt government announced that they would withdraw troops from Toulon for action in other theatres, since Mulgrave had matters so much under control.

While the number of French besiegers was increasing with the dispatch south of the troops victorious at Lyons, Hood's manpower shortages were becoming more acute by the day. When he quite reasonably asked the governor of Gibraltar to release 1500 troops there to make up for the withdrawals ordered by Pitt, arguing with impeccable logic that Gibraltar only ever needed to be garrisoned when there was a threat from Spain – but Spain was now an ally – the governor sent him just half the number. These 750 men arrived at the end of October together with a new general, Charles O'Hara, whom the War Office sent as a replacement for Lord Mulgrave. Just when Hood was consoling himself on the 'half a loaf' principle, further orders came from London ordering an entire regiment to leave Toulon for Gibraltar, there to be embarked for service in the West Indies. Hood and O'Hara were forced into the embarrassing situation of having to admit that all Lord Mulgrave's boastful talk had been mere bombast, and that the situation in Toulon was fast deteriorating.

By now wherever Hood turned, he saw things going from bad to worse. Allied morale was low with the troops poorly provisioned and the late autumn rains starting. Bread was a particular problem, with an overall shortage and maggoty loaves; the French had by now cut off the water supply to Toulon's flour mills. The Spanish were being more troublesome than ever: the king in Madrid promoted Gravina to lieutenant-general so that he would outrank Hood and thus automatically become allied commander-in-chief, but the British refused to accept this, and Anglo-Spanish diplomatic tensions mounted; as soon as there was a rightist regime in Paris, Spain would lose no time in switching sides. Hood hit back by promoting O'Hara to lieutenant-general also, even though he did not really care for the newcomer at a personal level, hating his pessimism and much preferring the 'can do' attitude of his predecessor, Lord Mulgrave. To cap all, the newly arrived Neapolitans under Marshal Forteguerri announced that theirs was an independent command, not under the orders of either Hood or O'Hara.

IDENTIFYING THE REAL TARGET

Meanwhile, Napoleon's whispering campaign against Carteaux and the support of the four political commissars finally bore fruit; even Paris could no longer understand what Carteaux was playing at or what his ultimate strategy was. On 23 October he was posted away to head the Army of Italy, and in came General Doppet, another overpromoted revolutionary general who had reached his present eminence solely through the ardour of his Jacobin beliefs; a former dentist, he allegedly could not stand the sight of blood. Yet clearly he had been appointed on a probationary 'get-tough-or-else' basis by the Robespierres and therefore went into action immediately. He began by probing all around the perimeter, everywhere from Fort Mulgrave to Cap Brun, but finally appeared to settle on Fort Malbousquet, which he attacked on 15 November. This sally proved to be a feint, and the real target quickly became obvious – Fort

Mulgrave, the objective that Napoleon had been urging for a full two months. Doppet's attack, with 4000 men in three waves, was by far the most determined yet and initially made headway, with the Spanish defenders being forced back to their second line of defence. General O'Hara landed and directed a counter-attack in person, backed up by gunboats from the Petite Rade. Soon the French were taking severe casualties and broke off the engagement, leaving behind 600 casualties; the allies had sustained no more than 90.

Napoleon had been promoted to major when Doppet arrived, and another incoming senior officer boosted his position immensely. This was none other than Baron Jean-Pierre du Teil, who had been the commanding officer when Napoleon was at Auxonne and marked the young Corsican down as 'one to note'; it was du Teil's brother who had written the artillery manual that Napoleon found so impressive. Since General du Teil, officially the new artillery supremo, was almost permanently ill at Toulon, he gave his young colleague a blank cheque. Saliceti, Gasparin, Barras, Fréron, du Teil – there seemed no end of influential backers Napoleon could harness to his cause. Always with keen antennae, he had previously befriended Doppet when the general lay ill at Perpignan, contrasting him favourably with Carteaux and even writing: 'Get rid of your blistering plasters, so that we may go and stick a few on the enemy's neck.'

Doppet repaid the compliment at Toulon, reporting on Napoleon's combination of intrepidity and tireless energy. 'Whenever I visited the posts during my stay with that army, I ever found him at his; if he needed a moment's rest, he would lie down on the ground, wrapped up in his cloak: he never left his batteries.' Yet none of this personal sentiment weighed with Napoleon when he saw in Doppet an obstacle both to his own glory and to the fall of Toulon. Once again he complained bitterly to Paris about his commander. He wrote directly to the Committee of Public Safety that he was being let down and, specifically, that he needed more artillery, which was being denied him. Once again the Bonaparte lobby did the trick: after a three-week tenure Doppet ceased to command the besieging army.

While the French awaited their new commander, Hood and O'Hara considered their options. O'Hara's tour of inspection left him more pessimistic than ever, but he shrewdly underlined the real weaknesses of the allied positions: the perimeter was too long for successful defence, the outposts were not adequately fortified, and the water supply was by now dire, with the outposts likely to be without the precious liquid if winter snows cut them off from Toulon. Above all, there was the abiding nightmare of manpower shortages. When 1500 more Neapolitans arrived at the end of October, Hood could count on a paper strength of 19,500 defenders (5000 in Toulon, the rest scattered in all the garrisons and forts), but because of sickness and poor hygiene only about 12,000 of these were true 'effectives'. The only glimmer of hope O'Hara could offer Hood was that, if the promised 5000 Austrian troops and 10,000 Sardinians ever arrived, it might then be possible to launch a major sweep to clear the French out of all the immediate environs of Toulon. Meanwhile, all Hood could do was gnaw on his nails with anxiety as enemy strength increased – by mid-November French forces were 35,000 strong. Frantic pleas to London produced simply another senior administrator – and Hood already had plenty of those. The new appointee, supposedly to oversee municipal affairs in Toulon, was Sir Gilbert Elliott, as over-optimistic as O'Hara had been over-pessimistic. Not the least of Hood's trials was trying to integrate the insights of this Jack Spratt couple. Elliott's first task was to keep the troublesome French royalists at arm's length. The Comte de Provence declared himself regent for the seven-year-old Louis XVII and expressed a wish to make Toulon a royalist stronghold; Pitt's instructions to Elliott were that he was at all costs to prevent this.

FOUNDATIONS OF A NOTABLE CLIQUE

The new French commander turned out to be General Jacques Coquille, who used the *nom de guerre* Dugommier, a career soldier and yet another Bonaparte admirer, who allowed Napoleon *carte blanche* and even promoted him to lieutenant-colonel. By this time Napoleon had gathered a notable clique around him, the core of his future military staff and including many men he would raise to the status of marshal of France when he became emperor eleven years later. One of his first friends was Androche Junot, an intelligent and witty man then a young sergeant from Burgundy. When Napoleon asked for a volunteer soldier with good handwriting, Junot stepped forward. Already impressed by the man's calligraphy and general attitude, Napoleon was one day dictating a dispatch when a cannonball from a British warship landed near by and sprayed Junot's writing paper with sand. 'Good,' said Junot, 'we won't need to blot this page.' This soldierly humour was exactly what Napoleon liked, and straightaway he appointed Junot to his general staff. Another find was a 29-year-old lieutenant-colonel named Claude-Victor Perrin, who became the future marshal victor. Other future marshals who attached themselves to Napoleon at Toulon were a 19-year-old captain named Auguste Marmont and a 23-year-old lieutenant, Louis-Gabriel Suchet. Yet perhaps Napoleon's most valuable friendships were those he struck up with the 25-year-old Louis Charles Desaix, who was to prove his greatest-ever general, and his closest friend, 21-year-old Gerard Christophe Duroc. In terms of the future Napoleonic empire, Toulon was for Napoleon and his young friends the equivalent of an education at Oxford or Cambridge (or Yale or Harvard).

The relentless artillery duel continued between the Royal Navy warships and Napoleon's batteries. Bit by bit Napoleon tightened the screw, with more and more batteries and better guns, both in the old locations and in new sites. He decided to feint first towards Fort Malbousquet, laying down an intensive barrage to mask his real objective, which was always the Balaguier/L'Aiguillette/ Fort Mulgrave complex. From the night of 27–28 November Malbousquet was under fire from three batteries: a new one, named the Convention, hidden in an olive grove, and the 14 guns from the Poudrière and Farinière batteries, the former of which now possessed one of Napoleon's prized long cannons that could not only devastate Malbousquet but even reach the town of Toulon itself. The technical superiority of his gunnery was already evident, for the response from the allies was an inaccurate bombardment from Fort Mulgrave and the warships. Napoleon's only problem now was the approach of winter. With 35,000 mouths to feed, Dugommier's army was itself suffering a crisis of morale, with inadequate food (because of the allied blockade) at the very moment the icy winds of winter were beginning to bite. So bad was the supply situation that Barras and Fréron recommended to the Committee of Public Safety that Dugommier pull out and go into winter quarters in Vaucluse, ready for another attempt in the spring. The Robespierres, however, were not yet ready to admit defeat. They sent four fireships to harass the British fleet from the seaward side – not a crucial development, but an annoying diversion none the less. Hood became increasingly frustrated that the Royal Navy could not interdict seaborne supplies to the French, even though he knew rationally that his ships could not patrol every creek and inlet in the Mediterranean. But it was an open secret that Napoleon was being supplied by sea with his gunpowder and, inevitably, a search for scapegoats was opened. The Spanish, always only reluctant allies of Britain because they both embraced the monarchical principle, were thought to be in secret negotiations with republican France, and the canard did the rounds that they always offered only token resistance to French troops before abandoning their positions.

The coalition's greatest single danger

By now Hood and his captains had identified Napoleon and his artillery as the greatest single danger they faced. The danger was more acute than they knew for Dugommier had finally given Napoleon the *carte blanche* for which he had yearned ever since coming to Toulon. On 25 November Dugommier wrote to the Committee of Public Safety that 'he did not think he could submit a more luminous and more feasible plan than the one submitted by the major commanding the battery [i.e. Napoleon], that, having followed the ideas of that plan, he had in turn hurriedly drawn one up, and this plan, for which it was a pleasure to give its primary author credit, he laid before the council.' Dugommier even employed Napoleon's own mathematical phraseology, referring to 'an exact calculation of means ... their right proportions ... their respective bearings'; this was the language the young man habitually used when explaining tactics. Now all reports from Toulon pointed in the same direction and praise for Napoleon had become unanimous, with no dissenting voices whatever. In addition to the paeans from Dugommier, Saliceti, Gasparin, Fréron and Barras, du Teil weighed in with a eulogy for his protégé: 'I cannot find words to portray the merit of Bonaparte: a considerable amount of science, just as much intelligence and too much bravery, such is a feeble outline of the virtues of that officer.' Napoleon's importance to the Jacobins had meanwhile moved on a notch with the appointment of Lazare Carnot as *de facto* war minister, the member of the Committee of Public Safety responsible for directing all 14 of France's armies. Saliceti enjoyed particularly good relations with Carnot and was most persuasive. Carnot perused all the other plans for the siege on Toulon laid before him and dismissed them brusquely in favour of young Bonaparte's.

Finally Hood and O'Hara had made the preparations they needed and, at 4 a.m. on 29 November they launched a large allied force against Napoleon, hoping to sweep away his batteries once and for all. The attack began brilliantly, the French were taken by surprise, and soon allied troops were swarming over the gun emplacements: in quick succession they captured the Convention, Poudrière and Farinière batteries. But discipline then crumbled. Elated by their early success, the allies split into smaller, fragmented bodies to pursue and loot. It did not take Dugommier long to realize that he actually had a clear numerical superiority, and then he counter-attacked. The combat that resulted was frenzied and bloody, with desperate hand-to-hand fighting and the expenditure of some 500,000 cartridges. In one of the keenly contested mini-battles O'Hara himself was wounded and captured, hauled away for a long captivity in France. Almost unbelievably, it transpired that the Spanish troops in the vanguard of the attack on the emplacements had not brought any studs or other tools to spike the French guns; hurriedly they sent to Fort Malbousquet for the equipment, but it was too late. Dugommier swooped down and rolled back the allies with his reserves, driving the enemy back to Malbousquet. Outside the fort the allies made a stand, reinforced from troops within the stronghold. The French made three separate spirited attacks but were thrown back on each occasion. Gradually the advantage swung back to the allies as they received reinforcements, from the Spanish reserve and from the Moulins and Petit Saint-Antoine batteries. After seven hours non-stop fighting, both sides fell back exhausted and retreated to their original positions. Casualties had been very high by the standards hitherto obtaining at Toulon: the British lost 148 dead, wounded and missing, the Spanish 119, the Neapolitans 65 and the Sardinians a similar number. Dugommier reported 271 casualties. Yet he was the moral victor. This battle was the 'tipping point' when the French decisively began to gain the upper hand, and in his report Dugommier once more singled out Napoleon and the artillery.

THE FIGHT FOR FORT MULGRAVE

One reason for allied decline was the now desperate shortage of manpower suffered by Admiral Hood. Even as the last batch of 500 Neapolitans arrived, this was offset by the loss of the 300 men of the 30th regiment, his best unit. In addition, the people of Toulon itself were despairing of a successful outcome and growing disaffected. They blamed the British and their allies in Toulon's General Committee for just about everything: not allowing the Comte de Provence to come to them (though how that was supposed to help them was unclear), requisitioning materiel, introducing disease, profiteering, hoarding and general inflation. The General Committee responded to the harsh criticism by issuing decrees restricting freedom of movement, while Sir Gilbert Elliott declared that the burghers of Toulon had had it too soft and been pampered; he recommended an immediate draft of all the non-combatants and 'idle mouths' among the male population. But not everything was working in favour of the besieging French. It was clear to everyone that Dugommier and Lapoype would have to combine for a final push to take Toulon or disperse for the winter; troop numbers were up to 40,000 by now but food was running short. Yet getting the inert Lapoype to collaborate with Dugommier was not easy. Once again Napoleon had to appeal directly to the Committee of Public Safety in Paris to force Lapoype to agree to his plans; Dugommier was no problem, as he was prepared to rubberstamp everything young Bonaparte did. Napoleon proposed that Lapoype be given direct orders to make an all-out attack on the Faron massif as a classic diversion while Dugommier's infantry pinned down the garrison at Fort Malbousquet, preventing them from making sorties. He himself meanwhile would deliver the *coup de grâce* at Fort Malbousquet with his artillery.

Having received the green light from Paris, Dugommier called a meeting of his 11-man War Council, whose members included

Lapoype, General du Teil and Napoleon. Napoleon's three-pronged attack on Faron, Malbousquet and Mulgrave was approved, and the further refinement was added that there would be a long-range bombardment of Cap Brun, to sow maximum confusion. Early December saw Napoleon at his most energetic. By the sixth of the month he had the three batteries of Convention, Poudrière and Farinière at concert pitch. He then played his ace, which was to construct three more batteries shoulder to shoulder to the south of Fort Mulgrave, within mortar fire of the fort but capable of delivering the most devastating broadside in unison. The new batteries were given the names Jacobins, Chasse-Coquins (Nail the Buggers) and Hommes sans Peur (Fearless Men) and were built under the very noses of the Mulgrave redoubt – this was typical of Napoleon who had earlier constructed the Brégaillon battery in a single night practically under the bows of allied warships. A new artillery duel began, dwarfing anything that had gone before. A fusillade of 20,000 cannon balls a day was sustained by the Sans Culottes battery, and even greater punishment by Hommes sans Peur. This was muzzle-to-muzzle, close-up warfare between the big guns; by so shortening the range Napoleon was able to increase by a hundredfold the destructive power of his guns, but he was taking a huge risk as his batteries could well be annihilated. He tried to soften up Fort Mulgrave by a bombardment so intense that several mortar bombs could sometimes be observed in flight simultaneously. The allies retaliated in kind: the Hommes sans Peur battery received the fire from ten warships, two hulks and two bomb-ketches and on the first day nearly all its gunners were killed or wounded. No one could be found to take their places until Napoleon, displaying the brilliance at human psychology for which he became renowned, announced that this battery would henceforth be known as the élite unit, to which only the highest calibre warriors would be admitted; after that, everyone wanted to join and the recruitment problem was solved.

TOWARDS THE TURNING POINT

It was typical of Napoleon to have estimated enemy resources precisely. He knew that Fort Mulgrave had three main redoubts, revetted
in earth and timber, up to 4.5 meters (15 feet) high and surrounded
by deep ditches in which the defenders could protect each other
with crossfire; moreover, the ground between the redoubts was
also covered by fire from the earthworks. From behind protective
walls and firing through embrasures were 25 guns, a mixture of
32-, 12- and 8-pounders, with mortars in support and fronted by a
deep ditch. On paper the positions were impregnable. But in fact the
position was relatively weak on the French right, as Napoleon's
superb eye for terrain had noted. Although Napoleon began with
local inferiority in numbers – it was to become a hallmark of his
military style that he would go over on to the offensive even though
the weight of numbers favoured the enemy – the French had impressive depth in reserves so that if they could gain a toehold in
Mulgrave, more and more pressure could be applied. At this juncture the coalition had a nominal strength of 18,700 troops – 7000
Spanish, 2000 British, 6200 Neapolitans, 2000 Sardinians and 1500
royalist French – but illness had reduced this to about 11,000 effectives only, who had to man a perimeter 24 kilometres (15 miles)
long; to shorten it even fractionally would bring the British fleet and
the town of Toulon itself within devastating range of Napoleon's batteries. Against this the French had no less than 38,000 effectives,
including 1650 artillerymen. But the winter cold was biting and
Dugommier was secretly very worried about discipline and desertion in his own ranks. The critical encounter had to be staged very
soon or the French army might implode.

On 14 December, in dark, wet and windy weather, Napoleon's 11
batteries opened up on forts Malbousquet and Mulgrave at ranges
between 230 and 1800 metres (250 and 2000 yards); the idea was

that the cannonade would be doubled on the 15th and quadrupled on the 16th. Everything went to plan except that Napoleon himself narrowly escaped death on the last day when he was knocked off his feet by the wind from a passing cannonball. But the stormy weather worried the faint-hearted, so that when Napoleon told Dugommier he had weakened the forts sufficiently for the *coup de grâce* to be delivered, the general and his advisers hesitated. Heavy rainfall and lowering clouds cast a general gloom, and there were fears that the not highly motivated troops would not be able to adjust their musketry to deal with the weather conditions. Dugommier was in a quandary: he did not want to preside over a defeat but, if he drew back, the political commissars might report to Paris that the general's heart was not in the job; in the atmosphere of terror in France 1793 commanders had been executed for less. The irony was that it was the selfsame commissars who were expressing doubts about the impact of storms on the battle; they were placing each-way bets, shifting the blame in case the venture miscarried but forcing the luckless Dugommier to make the hard decisions. The general tried to duck responsibility himself by suggesting that Napoleon lead the attack, but the younger man quickly talked him round to heading the assault himself, arguing that the weather factor was irrelevant: artillery and bayonets would do the job. His clinching argument was that if the attack were called off, the huge amount of powder and ammunition already spent by the artillery batteries on the three-day cannonade would have been wasted, and there was no reserve supply for a second such attempt.

Finally persuaded, Dugommier prepared to lead his 5000 men through the gloom of a winter night. The attack began at 1 a.m. on 17 December, using the cover of darkness. One column advanced to assail Mulgrave head-on while another followed the coast to come in on it from the left and rear, cutting its communications by water. After an initial 'friendly fire' incident caused by the limited visibility of the storm, the French reached the outer lines of Mulgrave and ferocious hand-to-hand fighting began; almost to a man the first wave died at bayonet point. When this attack was repulsed, Dugommier sent in his reserve, headed by Napoleon and Colonel Mouret. More seasoned troops than the first wave, they fought with revolutionary *élan* and fury. Parapets were scaled, defenders bayoneted, there was more hand-to-hand fighting inside the fort; no quarter was asked or given. French fighting spirit was given a fillip by gross errors by the Spanish defenders, who did not move their reserves up from the strongholds of Balaguier and L'Aiguillette. The Spanish commander, Brigadier Izquierdo, tried to shift the blame for this mistake on Neapolitan defenders retreating from Mulgrave who, he claimed, caused chaos among his own troops. Certainly the Neapolitans did panic, and it is undeniable that the multinational force defending Mulgrave was a factor in its loss. After three hours of bloody, grim and ferocious fighting Fort Mulgrave was in French hands. Saliceti and Gasparin arrived after the fighting to give the revolutionary troops their political blessing and found their favourite, lieutenant-colonel Bonaparte lying wounded on the ground, having taken an English sergeant's pike in the inner left thigh just above the knee. He had his horse shot from under him and then, fighting on foot, been run through with a savage thrust. At first there was consternation, gangrene was suspected and an amputation thought necessary, but a military surgeon brought in for a second opinion pronounced the wound not serious. Ever afterwards Napoleon bore a deep scar.

TOULON'S FALL TO THE FRENCH

With Fort Mulgrave gone and 'little Gibraltar' a smoking ruin, the position of Balaguier and L'Aiguillette was clearly hopeless. Neapolitan panic had prevented the reinforcement of Mulgrave and now it destroyed whatever slender prospects there might have been for the two strongholds on the promontory; the men of Naples were reported running backwards and forwards, those on the promontory colliding with those fleeing from Mulgrave, ratcheting up the chaos another notch. The finest achievements on this dismal night for allied arms were of the Royal Navy, which managed to evacuate 2500 men from the promontory. Some 300 Britons and 70 Spaniards died that night, and there were a further 250 officially posted missing and 400 prisoners; the French counted no more than 80 dead. Malbousquet had meanwhile been kept busy by an elaborate show of force and a continuous artillery bombardment. The third phase of the operation, the synchronized attack on the Faron ridge by Lapoype, at first went badly. The French attackers at the eastern end were repelled three times, and no amount of revolutionary exhortation from the commissars Barras and Fréron could rally them. It was fortunate for the French that the assault on the western side of the ridge went well so that the garrison on the east was overwhelmed. The French force then united against Fort de la Croix Faron and, after a robust defence, this too fell. Swarming down from the heights on to the lower Fort Faron, the victorious revolutionaries appeared in such numbers that the garrison at Faron made only token resistance. By 4 a.m. the *tricolore* was flying both over Fort Faron and Fort Mulgrave. The victory was as complete as Napoleon had predicted.

A gloomy Admiral Hood summoned an emergency council to see if further resistance were feasible. The consensus was unanimous that the Royal Navy warships could no longer use either the Petite Rade or the port. Concern was expressed that all the fight had gone out of the Neapolitans, that morale elsewhere was poor, and that the long-promised Austrians would clearly not be arriving in time. The Neapolitans confirmed all the bad opinions by pulling out of Cap Brun, Saint-Mandier and Fort Missiesy without even staying to try conclusions with the French; some thought this meant they had secret prior orders to cease fighting once events turned away from the allies. As the revolutionary armies gradually occupied all key positions around the bay, people in Toulon itself, remembering the fate of Marseilles and Lyons, began to panic. Many flocked with their belongings to the quayside, begging for evacuation, desperate to avoid the bloody purge they could see coming. Hood, though, had more pressing problems of his own. He ordered all Royal Navy warships in the Petite Rade to shift their berths to the outer anchorage. But now, to general stupefaction, Hood was forced to admit that he had no general contingency plans for the French fleet – supposedly the great prize that had justified the occupation of Toulon in the first place. Everything had been predicated on success: the Jacobins were thought to be easily defeatable, after which Louis XVII would ascend the throne and Hood would hand back the Toulon fleet, which he had held as trustee, to the grateful plaudits of the French nation. A definite message from the Admiralty, stating that in the worst-case scenario, he should destroy the fleet, reached him only after Toulon's fall. Yet, as an intelligent man, Hood knew where his duty lay. He had been derelict in not thinking things through or ordering any coherent contingency plans in the event he was forced to evacuate. Evidently he had played the usual game of the military second-rater, simply waiting to see what would turn up and postponing hard and unpleasant decisions.

'MUCH BLOOD, BUT FOR HUMANITY AND DUTY'

Very late in the day, Sir Sidney Smith, a 29-year-old naval captain and later to be Napoleon's nemesis in Egypt, was given the task of destroying the Toulon fleet and given a single fireship with which to do the job. In the circumstances Smith performed brilliantly and managed to set the arsenal on fire on the evening of the 18th; the terrific explosion at around 9 p.m. appealed to Napoleon's romantic soul, as he himself conceded. Smith managed to set on fire 11 French ships of the line and four frigates but was unable to do more. Inevitably, the British were criticized for concentrating on this job of destruction and leaving the refugees of Toulon to their fate, and it is true that the 7000 or so Toulon citizens who were evacuated were taken off mainly on Spanish or Neapolitan ships or private craft. The heartless Smith had nothing but contempt for the refugees: 'They crowded to the water like the herd of swine that ran furiously into the sea possessed by the devil,' he wrote. Moreover, claimed the British, retaining Fort La Malgue until the last moment enabled them to embark 9000 troops. By dawn on the 19th all allied troops had embarked. The French, moving cautiously at first into Toulon as they had expected stiff resistance, confessed themselves surprised by the precipitate British withdrawal and took full propaganda advantage of it. Then they turned to take revenge on Toulon. The Committee of Public Safety had had a particularly bad fright in southern France and reacted with the vindictive reflex common after such fear. Mass executions began on 20 December: first 200 naval officers and seamen then, next day, 200 civilian 'collaborators'. Joseph Fouché, later Napoleon's police chief and an expert in exemplary terror, expressed the Jacobin credo: 'We are shedding much blood, but for humanity and duty.' Napoleon's enemies later tried to make black propaganda out of the Toulon massacres that continued into the new year (altogether over 2000 were massacred), alleging he was a moving force in the atrocities. He was actually appalled, as was Dugommier, but both knew better than to protest at people's justice. Napoleon's later claim that only counter-revolutionary ringleaders were shot at Toulon is humbug, but so too is Sidney Smith's outrageous and ludicrous claim that Napoleon personally mowed down hundreds of citizens.

All those on the victorious side at Toulon went on to bigger, if not necessarily better, things. Dugommier asked to retire on the grounds that he had successfully completed his mission, but was instead made commander-in-chief of the army of the Pyrénées-Orientales. Surprisingly, in view of the fact that the siege of Toulon was one of the Jacobins' finest hours, marking the last conceivable occasion when foreign intervention could have halted the French Revolution in its tracks and the extinction of Girondist hopes for nationwide civil war, most of those associated with this famous success survived the famous right-wing *coup d'état* of Thermidor (July 1794), which ended with the execution of more than a hundred Jacobin leaders, including Robespierre. Fréron, in December 1793 a fire-eating left-ist who wanted to raze Toulon to the ground, perhaps remembering the old saw quoted by the writer Tobias Smollett in 1766 that 'the king of France is greater at Toulon than at Versailles' (because of the fleet, the arsenal and the dockyards), next emerged clearly into day-light as the lover of Pauline Bonaparte a few years later. He had clung to the coattails of Barras, a master of chicanery who had spectacu-larly reinvented himself after Thermidor and would go on to head the bourgeois government of the Directory in the late 1790s.

But then all the consequences of Toulon seemed to run counter to reason and logic. Given that the British and their allies controlled both bays in Toulon for three months, the action there should have been a catastrophe for France instead of merely a severe set-back. Instead of destroying the Toulon fleet Hood dithered, with the result that 13 ships of the line and 5 frigates survived; although Sidney Smith had torched 11 warships out of 17, only 6 were totally destroyed and 5 were quickly repaired. Had Hood destroyed the entire fleet as he should have, there could have been no invasion of Egypt by Napoleon in 1798 and therefore, conceivably, no *coup d'état* in 1799 or empire thereafter. So far from being punished or dis-graced for his poor showing – surprisingly, when one considers the

Admiralty's treatment of Admiral Byng in 1757 – Hood was actually awarded over £265,000 in prize money for the French ships actually destroyed, of which the admiral received one-eighth – and this even though by maritime law and convention no prize money was payable for ships allegedly held in trust. It will be remembered that Hood defended his failure to destroy the fleet initially by claiming he was holding them in trust for Louis XVII. The favouritism of the Admiralty was as blatant as their lack of logic.

EMPEROR OF FRANCE, MASTER OF EUROPE

Naturally, for Napoleon Toulon was the first step on a career ladder that would soon have all the velocity of a moving escalator, though he too had to navigate the treacherous political shoals through the French Revolution, Thermidor most notably. On 22 December the assembled commissars promoted him to brigadier-general (he was still only 24) – a promotion confirmed soon afterwards by the Committee of Public Safety. All those who had been with him at Toulon rose alongside him – Desaix, Duroc, Junot, Marmont, Victor, Suchet – and Napoleon always had sentimental feelings about those who had served with him in the theatre where he first made his mark; even the useless Carteaux later became the beneficiary of his largesse, and he was no friend of Napoleon in 1793, to put it mildly. Napoleon's achievement at Toulon was not to spot that the western promontories were the key to the whole puzzle – that was obvious to most observers – but to work out a way they could be taken when everyone else was self-confessedly clueless. Demonstrating a patience that would desert him in the latter stages of his career, he had first fought the torpor, inertia and defeatism of others, demonstrating steely willpower, and had then meticulously built up the artillery to the point where it had local advantage over the enemy; it has been well said that his batteries functioned like the long jabs of a superb tactical boxer. The superlatives that Dugommier, du Teil and Saliceti showered on him were more than merited, for he demonstrated a gallery of talents not usually found together: boldness, rapidity and tenacity. He also showed that he had a fine intellect, a mind that could encompass at once the smallest details of military operations and also the entire overall picture: most talented soldiers are good at one or the other but rarely both. He had already proved that he was a marvellous reader of ground and that he could get into the enemy's mind, empathize and then out-think him.

Naturally Toulon, as a self-contained operation somewhat limited in scope, did not reveal the mature genius of Napoleon as commander, but the mathematical principles he applied there remained consistent over the rest of his career. Here is his credo as expressed to one of his admirers:

> Military science consists in calculating all the chances accurately
> in the first place, and then in giving accident exactly, almost
> mathematically, its place in one's calculations. It is upon this point
> that one must not deceive oneself, and yet a decimal more or less
> may change all. Now this apportioning of accident and science
> cannot get into any head except that of a genius. Accident, hazard,
> chance, call it what you will, a mystery to ordinary minds, becomes
> a reality to superior men.

It is tempting to see Toulon as a kind of laboratory experiment for the methods of the mature Napoleon. The use of artillery as a key weapon would remain a constant in his thinking, and the murderous struggle for the Great Redoubt during the battle of Borodino in 1812 shows Napoleon still trying to overpower the enemy with the sheer quantity of shells and cannonballs expended; but the truth is that by that time his enemies had learnt the lessons of Toulon better than the master himself. The clever positioning of his forces between forts Malbousquet and Mulgrave reveals in embryonic form his later fondness for the 'centre position', interposing his army between two parts of the enemy's forces so as to destroy it piecemeal; Napoleon was always brilliant at achieving local superiority by concentration of force, even against a numerically superior foe. Finally, Toulon reveals clearly the importance Napoleon always attached to speed. One of his favourite ploys, evinced in campaign after campaign, was to station his forces two days' march away from a hostile force on, say, a Sunday, leaving the enemy to believe that battle would be joined on Tuesday; by forced marches overnight he would then appear on the field on Monday against an unprepared opponent. Some of the mileages achieved by Napoleon and his marshals speak of almost incredible speed and mobility. He himself covered 80 kilometres (50 miles) in three days during the Italian campaign of 1796–7, his marshal Soult and his corps covered 445 kilometres (275 miles) in 22 days during the Ulm campaign of 1805, while in the same year at Austerlitz another marshal, Davout, moved his army corps 140 kilometres (88 miles) in 48 hours.

Toulon was simply the beginning of Napoleon's remarkable career. Surviving the fall of Robespierre when the French Revolution swung right in July 1794 in the famous 'Thermidorean reaction', he clung for a while to the coat-tails of the supreme political operator Barras, his early patron. As Barras's hitman he dispersed the Paris mob in 1795 with the celebrated 'whiff of grapeshot'. But it was the Italian campaign of 1796–7 that first revealed his military genius in all its splendour: he won a string of victories and forced Austria to the conference table. By now his sights were set on supreme power but he judged the time was not quite ripe, so departed on an amazing 13-month campaign in Egypt (1798–9), winning more victories against more primitive armies. Returning to France, he found Barras and the other members of the five-man ruling council or Directory thoroughly discredited. Late in 1799 he masterminded a *coup d'état* that brought him to supreme power; for the time being he masked the full scope of his ambitions and accepted a position as first consul. Further successful warfare against Austria in 1800 ended with a general European peace (1801–3), but when hostilities predictably resumed, Napoleon tried hard to invade England; the ceaseless vigil by the Royal Navy under Nelson thwarted him. In 1804 he threw off the veil and had himself crowned emperor.

ON THE ROAD TO WATERLOO

The years 1805–7 saw Napoleon at the height of his powers. Ignoring the defiant British, he won a succession of battles, one in each year, against Austria, Prussia and Russia respectively. But from 1808 onwards his position as master of Europe began to unravel. First, he was sucked into a pointless war in Spain. Then he conducted a hard-fought campaign against Austria (1809), which he won narrowly on points. Yet it was clear to keen-eyed observers that the technological gap between France and its enemies had closed, and that there would be no more easy French victories. Next, when his attempt to bring Britain to its knees through an economic blockade failed, he blamed the Russians for sanctions-busting and made the greatest mistake of his career by invading Russia (1812). In a disastrous campaign he lost half a million men. Now forced on to the defensive, in 1813 he confronted a grand coalition of Britain, Prussia, Austria and Russia and was finally brought to bay at the decisive battle of Leipzig. When the victorious allies invaded France in early 1814, Napoleon was forced to abdicate and was exiled to the island of Elba. In 1815 he staged a return to the mainland, France rallied to him, and he fought his final campaign, the 'Hundred Days', which ended with his defeat at Waterloo by Wellington, the British, Dutch and Prussians. Exiled to

St Helena in the Atlantic, he died there in 1821 (almost certainly poisoned).

A final verdict on Napoleon as a commander is not easy, if only because so many moral considerations have to enter the equation when we balance the pluses and minuses. At least 4 million people died in Napoleon's wars, and the death tolls in his major battles remained unequalled until the slaughterous world wars of the twentieth century. The dead in his many battles have been calculated as follows: Austerlitz (1805), 16,000; Ulm (1805), 18,000; Jena (1806), 50,000; Friedland (1807), 28,000; Aspern (1809), 45,000; Wagram (1809), 74,000; Borodino (1812), 80,000–120,000; Lutzen (1813), 45,000; Bautzen (1813), 40,000; Dresden (1813), 50,000; Leipzig (1813), 90,000; Waterloo (1815), 22,000. These casualty rosters do not include the war dead in Italy (1796–7), Egypt (1798–9), the Austrian campaign of 1800, his campaign in France in 1814, the other battles of the Hundred Days (1815), or any of the battles fought by his marshals, especially those in Spain during five years of blood-letting in 1808–13. Morality aside, there are many other criticisms that can be made of Napoleon. He was not an original in warfare but simply refined existing methods; he introduced no new methods in tactics or waging war, though he did excel at the methods that already existed; after 1807 he was less impressive as warfare became more an affair of attrition than a battle of annihilation, and, by a kind of poetic justice, the increasing importance of artillery, which he had been the first to underline, meant that the efficient command and control structure and the corps system of several independent armies, which had served him so well in the glory years of 1805–7, no longer operated in his favour. Moreover, Napoleon was always at his best as a commander of small armies, peerless as a captain of forces 30,000–60,000 strong, which he could use in blitzkrieg warfare. The brilliant campaign he fought in France in the early months of 1814 saw him back to his best, mainly because he was once again leading small forces; huge, cumbersome armies always caused him logistical problems he could not overcome.

A PEERLESS GENERAL?

Let us, then, first play 'promoter of the faith', emphasizing Napoleon's great qualities, before becoming devil's advocate and pointing out his many deficiencies. These qualities may be summed up as charisma, intellect, workaholism and willpower. Napoleon clearly had a magnetic personal presence, which led even tough veterans and fire-eating generals to quail before him. He had that prerequisite of charm – making the person he was talking to feel that at that moment he or she was the most important person in the world. He had a remarkable ability to sway other men to his way of thinking and get them to fulfil his aims and purposes; women he either browbeat if he did not like them or seduced if he did. He worked hard to develop the common touch and pretended to know the personal histories of every man in his 200,000-strong army – an obvious absurdity, but its psychological effect was undoubted. Napoleon did know everything about his senior officers, and in the case of others he had his aides coach him in names and details beforehand. His intellectual powers were of a very high order, and he had the rare combination of mathematical talent and a superb memory. He also knew where every one of his units was, who commanded it, the personality of that commander, the unit's manpower and its weaknesses – and we are talking about seven army corps totalling 200,000 men. He could remember the last time he met someone – and he met thousands – and what the conversation had been about. He was not just a master of all the military arts but a talented lawmaker, a shrewd politician, an intellectual and an excellent psychologist. His energies were superhuman: he often worked 18–20 hours a day, even occasionally going three days without sleep. And his willpower was literally fabulous. He did not believe that politics was the art of the possible; he did not believe that everything was written in the deterministic sense, and clearly believed in the supremacy of mind over matter. 'Moral force rather than numbers decides victory' was one of his maxims, and he frequently declared that in warfare and human affairs in general the moral outranked the material in the ratio of three to one. As a general he was peerless, frequently achieving the destruction of the enemy in a single battle.

Napoleon's critics allege that his shortcomings outweigh these great talents. Although he was a shrewd psychologist in the sense of knowing what motivates human beings, he was actually a poor judge of individual human character. He was betrayed to the point of treason by four separate people, the marshals Bernadotte and Murat, the diplomat Talleyrand and his chief of police Fouché, yet never acted decisively against them. A true despot would have executed all four. If anything, he was insufficiently ruthless, as shown by his absurd indulgence of his four useless brothers. He managed the rare feat of meddling micromanaging while at the same time giving his incompetent marshals their head. His strategic limitations were glaring, for only a fool would have invaded Russia in 1812 while the 'Spanish ulcer' was still draining his resources in the Iberian peninsula. The most iconoclastic revisionism concerns his true status as a great commander. Critics say that his victories were mainly won against mediocrities and nonentities and that he imploded when faced with truly talented opposition. It is true that after 1800 he won only two clear victories, at Austerlitz in 1805 and Friedland in 1807. His general Louis Charles Desaix was the real victor at Marengo in 1800 and Marshal Davout was the true victor of Jena-Auerstadt in 1806. The victory at Wagram in 1809 was a narrow, on-points, makeshift and unconvincing affair, and the two dreadful battles against the Russians, at Eylau in 1807 and Borodino in 1812, were indecisive encounters marked by maximum bloodshed. Above all, Napoleon was unimpressive when it came to managing large armies and coordinating them across a number of fronts, as in the Russian campaign in 1812. He lacked the genius of the Mongol general Subudei, history's greatest strategist, who could coordinate multiple armies over multiple mountain ranges and use one army to screen another's flank. Certainly in percentage terms, Napoleon had nothing like the success rate of the truly great captains in history who had 100 per cent or near 100 per cent records: Alexander the Great, Hannibal, Julius Caesar, Subudei, Tamerlane, Turenne. Against this it can be urged that none of these great ones ever fought so many battles as Napoleon did and that they benefited, as he did not, from top-class armies already in existence. In Napoleon's favour it can be said that he never lost a single battle against a single adversary or against a single nation; the great defeats, like Leipzig and Waterloo, were always against a coalition of enemies.

The Warrior Mind

The psychology of the warriors

TO DESCRIBE THE FEATS of great warriors on the battlefield is one thing; to attempt to enter their minds and probe their psychology is quite another, and indisputably much more difficult. It would be simple if there were some one overarching psychological principle that would explain the mentality of the warrior, but there is not. All general explanations are ultimately found wanting. The psychologist Alfred Adler thought that the will to power was the basic drive in all humans, but did not explain why only a tiny minority of instinctual people were able to actualize that instinct, as warriors clearly could. The Swiss psychologist Carl Jung placed warriors in the category of 'extrovert–intuitive' in his investigation of psychological types, where he elaborated an eightfold typology – four types of introverts and four types of extroverts. But since explorers, big-game hunters and even stock-exchange speculators were also found in this group, any investigation into the psychology of the warrior was not greatly aided.

THE 'WARRIOR ARCHETYPE'

Much work has been done by modern Jungians on the so-called 'warrior archetype', but this consists in decanting the purely military qualities from the alleged warrior and seeing whether the basic archetype can assist those interested in a business or creative career. The type of book that specializes in leadership secrets – 'the leadership secrets of Attila the Hun,' say (such a book does exist), is along these lines – trying to detach the intellectual content of battle decisions from the bloody business of actually fighting. The founder of psychoanalysis, Sigmund Freud, implied that warriors were most likely to be favourite sons: 'A man who has been the indisputable favourite of his mother goes through life with the feeling of a conqueror.' This would certainly work well for Richard the Lionheart, clearly the apple of Eleanor of Aquitaine's eye, but not for Napoleon. The plain fact is that throughout history warriors have come from a variety of backgrounds: some have been parental favourites, others not; some eldest sons, others not. Any general attempt at categorization is doomed to failure.

This becomes very clear if we examine our chosen sextet of warriors. All promising generalizations soon founder. Adler speculated that individuals with 'organ inferiority' compensated with a superior lust for power, and so the great dictators were famously short of stature or otherwise physically impaired. In the twentieth century the examples of Stalin, Hitler, Mussolini and Franco are well known. In the case of our warriors, it is true that Napoleon, Attila and Cortés were all of below average stature, and that Tokugawa Ieyasu was 'corporeally challenged' by his obesity. On the other hand, Richard the Lionheart was a tall man and by all accounts a splendid physical specimen, and we can surely infer something similar for Spartacus, for he would not otherwise have survived so long in the arena.

The idea of being toughened by a disturbed childhood is also superficially attractive. Cortés, Spartacus and Attila may be assumed to have had a hard schooling in early life, and Tokugawa Ieyasu certainly had the same childhood memories as William the Conqueror – scarcely knowing from week to week if he would survive. Yet Napoleon and Richard the Lionheart had fairly easy childhoods. Maybe sexuality holds the key? It is sometimes said, and has been expressly stated in the case of Field-Marshal Montgomery, for example, that a warrior of necessity lives in a macho 'homosocial' world that would predispose him towards bisexuality. This would certainly hold good in the case of, say, Alexander the Great and Julius Caesar, and we know for certain that Tokugawa Ieyasu was bisexual. And it is surely interesting that charges of bisexuality have been laid at the door of Spartacus, Richard the Lionheart and Napoleon, admittedly on purely speculative grounds. But this idea too is decisively refuted by the careers of Attila and Cortés, rampant womanizers and polygamists both.

'ONE MURDER MAKES A VILLAIN, MILLIONS A HERO'

One obvious issue immediately presents itself. Are all great warriors basically psychopaths? Warfare is an area where success can only be bought with the blood of large numbers of human beings. It is difficult to assess the death tallies in the case of our sextet, especially in the ancient world, where all historians routinely added several noughts to every number. We know that 50,000 men fell in the battle of Sekigahara alone, but it is impossible to quantify the rest of the losses caused in the era of constant civil war in Japan. An educated guess would be that Spartacus's two-year rebellion cost 100,000 lives, and Attila's career maybe twice that. Richard the Lionheart spent a lifetime fighting so no overall estimate is possible, but the deaths on both sides during the Third Crusade must have amounted to 100,000. Cortés's conquest of Mexico is thought to have cost a quarter of a million lives, to say nothing of the devastation caused later by the diseases introduced by the conquistadores. As for Napoleon, the best estimate is that his twenty years of warfare were responsible for four million dead.

These are controversial matters, but the cluster of ideas around 'psychopathy' has been debated from the very earliest times. In the great Hindu epic *The Mahabharata*, the famous warrior Arjuna questions his motives in being able to slaughter so easily, and there are similar passages in *The Iliad*. Psychopathy in the normal sense is unlikely to provide a sufficient explanation for the conduct of warriors, if only because the stress of battle, experienced by leaders as well as the rank and file, is a constant in accounts of warfare. It is well known that the constant stress of battle produces a peculiar psychology that may even ultimately be susceptible to a chemico-physical explanation. By definition, a true psychopath would not experience such stress.

A more promising avenue opens up if we distinguish between the psychopath proper, devoid of all moral sense, and the sociopath, whose moral sense has been temporarily or permanently eliminated by existence in a hypermasculine environment. It has been speculated that there is a potential for moral breakdown in humans that can be triggered by natural disasters, social collapse or even in a football match if the other side wins. It is obvious that it is above all in war that such 'sociopathy' would manifest itself; after all, otherwise, how could soldiers ever be induced to kill? It is the oldest cliché in the book that what is heroism in wartime is murder in peacetime. In the words of the cynic: 'One murder makes a villain, millions a hero.'

WHAT IS A WAR CRIME?

The psychology of the warrior can be understood only if we put a bracket around normal peacetime morality. This inevitably leads us to another conundrum. What is a war crime? The cynic contends that war crimes are what the losers in a conflict have done; the atrocities of the winners are merely 'collateral damage'. On paper the bombing of Dresden and Hiroshima and Nagasaki in the Second World War were obvious crimes, but the defenders of these actions as 'necessary' like to stress the moral superiority of the cause for which their perpetrators were fighting. This is obviously a slippery slope, as we can see in the case of our own study. By a bizarre coincidence both Napoleon and Richard the Lionheart have been accused of war crimes, both in the Middle East and in almost exactly the same circumstances and geographical location. And in both cases the sinister doctrine of 'illegal combatants' was invoked. After his successful siege of Acre in 1191, Richard accepted an offer to have all 3000 Saracen prisoners ransomed for 200,000 dinars, the money to be paid in two instalments, and 50 per cent of the prisoners released when each tranche of money was paid. The disingenuous Saladin did not have the money and spun the negotiations out, hoping to find some pretext that would enable him to welsh on the deal. Finally, finding that Richard would not be duped into any of his traps, he mendaciously claimed that all the prisoners had to be released after the first instalment of the money, and a handful of hostages handed over pending the payment of the second one. Infuriated by Saladin's prevarication and dishonesty, Richard's patience snapped and on 20 August he ordered a mass execution of the captives. The crusaders set on the Muslim prisoners with lance and sword, and soon completed the task of mass butchery.

Richard alleged that there were a number of reasons for his action. He did not have enough food to feed them but could not leave such large numbers of heathen prisoners in his rear when he marched south, especially as they were men whose so-called word of honour meant nothing. He thought that Saladin would renege on the deal if he got all the prisoners back and would sacrifice his hand-picked hostages rather than pay the second tranche of the money. Moreover, massacre was an accepted part of warfare in the crusades: Saladin had butchered all the Hospitallers and Templars after the battle of Hattin. Finally, Richard was offering both his own men and the enemy incontrovertible proof of his own toughness. His credibility was on the line unless he called Saladin's bluff.

The comparison with Napoleon in 1799 is eerie. In February that year, as part of his Egyptian campaign, Napoleon besieged and took the cities of Gaza and Jaffa in Palestine – exactly where Richard had been 600 years earlier. Napoleon then ordered over 4000 prisoners massacred. Anxious to save bullets and gunpowder, Napoleon ordered his veterans to exterminate the captives with cold steel or by drowning. Napoleon used many of the same arguments as Richard. He did not have enough food to support his own men, let alone prisoners; the other side had already beheaded prisoners; those who had previously surrendered under parole rejoined the fighting once his back was turned, showing that their word of honour meant nothing, and, above all, his own credibility was involved.

It is interesting to weigh Napoleon and Richard in the moral scale for broadly similar crimes. Richard had at least one advantage Napoleon did not possess: by the standards of his day, slavery was not an egregious evil, so he could have sold off the prisoners Saladin refused to ransom instead of executing them. Also, Saladin must share some of the blame for the massacre, not just by his stalling but by his ruthless calculation as well. From his point of view, sacrificing 3000 people who were neither élite troops nor wealthy individuals might have made more sense than paying over 200,000 dinars; if he did not have that sum, he was dishonest to have struck a deal with Richard on that basis. Moreover, Richard, by his own lights, was engaged in a holy war against blasphemous unbelievers, and the Pope himself had sanctioned a 'no quarter' approach. Napoleon did not really have the option of selling his captives into slavery. On the other hand, he had supposedly invaded Egypt in the first place to bring 'civilization' to benighted barbarians, and nothing in contemporary western culture, either sacred or profane, sanctioned the massacre of prisoners. Napoleon, in short, had fewer excuses than Richard. Yet their common reaction in almost identical circumstances underlines the most obvious lesson about warriors: to be successful, they must be supremely ruthless. Cortés and his lieutenants, too, went in for the mass slaughter of civilians who were going about lawful and peaceful occasions. Spartacus practised crucifixion, and Attila impalement, while Tokugawa Ieyasu relished the mountain of skulls after Sekigahara.

THE ATTRIBUTES OF A WARRIOR

Comparisons between and within our sextet of warriors open up more promising lines of investigation than an attempt to find a one-size-fits-all type of overarching psychological theory of warriors. Our understanding of warfare and the mentality of the great (and not so great) captains of history is deepened by the number of interacting factors we can identify, and also by the complex meanings set up by both the similarities and differences between our subjects. Napoleon, Richard and Attila faced first-class military opposition at some stage in their careers, in the shape of Wellington, Saladin and Aetius respectively, whereas Cortés, Ieyasu and Spartacus did not. If we operate a moral scale, we would locate Spartacus at the top and Cortés at the bottom. If military genius is the touchstone, the descending rank order would be Napoleon, Richard,

Cortés, Attila, Ieyasu and Spartacus. Napoleon and Cortés fertilized forces of cross-cultural change in ways that the others did not. Only Ieyasu can be said to have fought non-ideological battles. The possible modes of comparison are multitudinous. Although there is no one psychological type or mindset that allows us to differentiate warriors in the true sense from other adventurers – for instance, the explorer H. M. Stanley, the man who found Livingstone in central Africa, had the mentality of a warrior even though he never commanded large armies – certain common traits can be clearly distinguished.

All successful warriors should be monomaniacal practitioners of the art of war, should have superhuman energy and should start young on their careers. To be great, one needs to have a range of rare qualities: to be able to adapt and improvise, to be able to exploit terrain and weather, to be able to read the enemy's mind, to have skill in winning over allies, to be at once decisive and flexible, to understand the importance of morale, to possess keen political understanding, to be able to identify simple solutions to problems, to know how and when to strike at the enemy's Achilles' heel, to be able to think through the circumstances of victory even before battle is joined and, crucially, to be lucky.

SPARTACUS: THE CHARISMATIC GUERRILLA WARRIOR

In the end, though, only a case-by-case analysis will enable us to get even to first base when it comes to entering the minds of our sextet of warriors. In the case of Spartacus, our problems are particularly acute. An illiterate gladiator, he left no famous 'slave literature' like the writings of Elaudo Equiano in the eighteenth century or Frederick Douglas in the nineteenth. Everything we know about him we know from his enemies, and they do not even, as with Attila, provide us with a close-up portrait of him. It cannot be too strongly emphasized that we have to distinguish the Spartacus of history from the 'Spartacus' of Howard Fast, Arthur Koestler, Kirk Douglas or Aram Khatchaturian. Here, for example, is a typical nineteenth-century attempt to imagine Spartacus, taken from a series of didactic monologues by E. Kellogg, widely used at one time to encourage the art of public speaking. The author imagines Spartacus addressing his fellow slaves at Capua.

> If ye are beasts, then stand here like fat oxen, waiting for the
> butcher's knife! If ye are men, follow me. Strike down yon guard,
> gain the mountain passes and there do bloody work, as did
> your sires at old Thermopylae! Is Sparta dead? Is the old Grecian
> spirit frozen in your veins, that you do crouch and cower like
> a belaboured hound beneath his master's lash? O comrades!
> Warriors! Thracians! If we must fight, let us fight for ourselves.
> If we must slaughter, let us slaughter our oppressors! If we must
> die, let it be under the clear blue sky, by the bright waters, in
> noble honourable battle!

With Spartacus, above all, we must infer mental states from battlefield deeds. His actions clearly align him with history's noble lost causes, whether we think of William Wallace in Scotland, the abolitionist John Brown at Harper's Ferry, Che Guevara in Bolivia, the rebels in the 1871 Paris Commune or even the Ibos in the Nigerian Civil War. There is no evidence that Spartacus was in any sense a revolutionary and his brutality was as routine and unthinking as the Romans'. Did he have a great strategic vision and was his project to cross the Alps and escape Roman tyranny thwarted only by the greed and short-sightedness of his followers? We have no means of knowing, though circumstantial evidence suggests he was always riding a tiger, in many ways as much at risk from his booty-crazed confrères as from the Roman armies. The march on Rome seems to have been a mere stratagem to keep his forces from imploding, though the ban on precious metals hints at something more deep-thinking.

The idea of Spartacus as a great general appears to be merely exaggerated Marxist rhetoric, though a case can be made that he carried out two of the precepts later made famous by Mao and his handbook on guerrilla warfare. Mao wrote that the peasantry was the sea in which guerrillas needed to swim, and Spartacus certainly seems to have understood this intuitively, even if Roman tyranny would have made most inhabitants of the Italian countryside willing helpers for the Spartacists. In his campaigning, too, Spartacus seems to have anticipated Mao's famous advice to an irregular army: 'When the enemy advances, we withdraw/When he stops, we harass him/When he tires, we strike/When he retreats, we pursue.'

ATTILA THE HUN: THE WARRIOR AS MAFIA BOSS

Spartacus is peculiarly opaque because of the lack of sources, but many of the same problems await us in the case of the other illiterate warrior, Attila. Here Priscus's embassy is the vital, and almost the sole, clue. The ambition and ruthlessness are plain to see, as also the shrewdness at reading individual character face to face, but the two principal weaknesses in Attila's mental armour seem to have been vaulting ambition and the lack of political antennae. In his case self-confidence shaded over into overweening arrogance, and he seems genuinely to have thought that he could become master of the known world. His envoys regularly trumpeted that their master, if he wished, could own every blade of grass in Europe and Asia. As Priscus reported at the time of his embassy: 'No previous ruler of Scythia ... had ever achieved so much in so short a time.

He ruled the islands of the Ocean [sc. the Atlantic] and, in addition to the whole of Scythia, forced the Romans to pay tribute ... and, in order to increase his empire further, he now wanted to attack the Persians.'

Attila is a classic example of what happens to a small talent when success goes to its head. He had engineered a brilliant protection racket that for a time kept his position unassailable, but he never seems to have grasped that, without the hundreds of thousands of whooping Hun warriors Christian propagandists falsely alleged were at his disposal, his power base was always precarious because of the manpower problem. As a diplomat Attila was a disaster, as shown clearly in the western campaign of 451. Even a mediocre manipulator should have been able to drive a wedge between Aetius and his eventual allies in France, which would have allowed the Huns to fight their enemies piecemeal. But Attila's foaming-mouthed oratory about Honoria convinced many that he had taken leave of his senses, that he was not a man with whom one could do rational business. He managed to unite Aetius and the Visigoths and alienate his putative ally Geiseric at the same time. Finally, we should mention that, although Attila has often been hailed as a forerunner of the Mongols, he did not create an empire from scratch as Genghis Khan did, but received his dominion ready-made from Rua and Octar. In this respect he was analogous to Alexander the Great who inherited a formidable army from his father, Philip. Attila, in short, has been overrated because of overwrought Christian propaganda about the 'wrath of God'.

In the spin put on his achievements by Christian writers, we can perceive a curious similarity between Attila and Cortés, both men with a clear psychopathic streak. There is some evidence that the later Attila, of the era 451–2, is a degenerate version of an earlier, more clear-headed and austere warrior. When Priscus was on his embassy with Maximius, he was impressed by the spartan and strenuous quality of the Huns. One of them told him the reason the Huns were superior to the Romans was that among the Huns every man bore arms and fought for his interests, whereas among the Romans, who had grown soft and effete, the rich and powerful expected others to do their fighting for them. When Priscus objected that the Huns had to be on a permanent war footing and could never enjoy the blessings of peace, Onegesius's freedman replied that peacetime simply meant heavy taxation and the depredation of criminals, against whom, the 'civilized' authorities would plaintively whine, they could do nothing. The so-called 'evils' of war, on the other hand, meant (though the Huns would not have used such words) the morality of strenuousness, the frontier ethos, the idea of every man being able to defend himself. Moreover, the Huns were content with their economic lot; it was not like Rome or Constantinople, where the highly privileged possessor of a fortune of 1 million sesterces still continued to cast envious eyes on those richer than himself. Yet by 451, Attila was forcing a change of mentality on his followers. In order to persuade them into the quixotic invasion of France, carried out for Geiseric's objective interests, not theirs, he had in effect to corrupt his warriors, who lived austerely, by promising them previously unheard-of levels of wealth. By this time, it was reported, Attila would no longer listen to reason but was increasingly in a world of his own imagining.

Certainly the Attila of 451, with his insane demands to have Honoria delivered up to him, was no longer the calculating blackmailer of yore. Perhaps this tendency was always latent in him, for Priscus noticed a disconcerting impatience with inconvenient details. When the proposal to invade the Persian empire was mentioned, one of the Byzantine envoys asked how Attila could convey an army from central Europe to Persia. Attila answered that if you followed the north coast of the Black Sea all the way to the end, you would arrive in Persian domains without even having to set foot on Byzantine territory. He seemed to have no idea of the immense distances involved and appeared to be working from a kind of collective memory of the Huns' trek in 395–6. The obvious objection, of course, was that the Huns were then living north of the Black Sea, so an incursion into Persia made sense, as it no longer did. Attila seemed to be contemplating an adventure that was almost wholly a leap in the dark. Ambitious plans were being drawn up without proper maps or supply routes, purely on the basis of some vaguely remembered geography. Lust to conquer the known world is one thing, but at least Alexander and Genghis Khan went about their ambitions in a rational and scientific way. The mind of Attila was encumbered by (maybe quite literally) castles in Spain (where his enemies the Visigoths were).

During the invasion of France, Attila hoped to live off the land and made no attempts to construct a credible supply chain from Hungary to central Gaul. That was why, had Aetius so desired, the Huns could have been annihilated after their defeat at Châlons. Attila's impatience with logistics means it is impossible to consider him as one of the great captains of history. As a much shrewder strategist, Frederick the Great, was later to observe: 'Without supplies, no army is brave.' Poor at logistics, poor at grand strategy, Attila on close inspection increasingly looks like a military one-trick pony, heavily dependent on the shock impact of his mounted archers, and with no shots in his locker if the Hunnic arrows failed. He had made a good career as an extortionist and should have stuck to demanding money with menaces. Essentially a large-scale mafia boss, Attila had the good fortune to appear at a historical moment when the eastern half of the Roman empire was very weak and the western half in terminal decline.

TOKUGAWA IEYASU: THE WARRIOR AS CHESS-PLAYER

The most interesting psychological profile in our sextet of warriors is that of Tokugawa Ieyasu, but there is much about his cast of mind that must remain obscure, to a westerner at least. All great men are of course to some extent the products of their culture, and there is much about the culture of the Era of the Warring States that is opaque or elusive. It is difficult, for example, fully to understand the culturally determined cult of *shudo* – that age-structured system of bisexuality whereby older men teamed up with boys concurrently with their relationships with women. Sometimes the relationship ended when the boy came of age, but sometimes it continued into adulthood, as with Ieyasu and Ii Naomasa, leader of the Red Devils at Sekigahara. Naomasa was 39 at the time and hailed from the Ii clan of Totomi province, particularly favoured by Ieyasu. Badly wounded by a sniper at Sekigahara, he lingered on for another 18 months, lived to see his lover's final triumph but finally died of his wounds in 1602.

But the Ieyasu–Naomasa liaison was only one of many such famous partnerships. Oda Nobunaga had the teenage lover Mori Ranmaru while Takeda Shingen, Ieyasu's early military nemesis, had Kosa Ka Masanobu. When Toyotomi Hidetsugu committed seppuku in 1595 he was joined by his favourite, Fuwa Bansaku, while the man who was shogun until 1565, Ashikaga Yoshiteru, had two famous partners, Matsui Sadononokami and Odachidono. The difficulty for the historian lies in deciding what, if any, influence the culture of bisexuality had on general modes of thought and military decision-making. To what extent was Ieyasu's an original mind, and to what extent can it be seen as an obvious product of the culture of the age? The same question arises with Alexander the Great and his famous relationship with Hephaistion. Since both Alexander and Ieyasu live on as 'great men' for the ages, we must presumably conclude that their cultural milieu was only marginally significant.

Even without a profound understanding of Japanese culture, one can draw certain incontrovertible conclusions about Ieyasu. He was supremely ruthless. He killed Hideyoshi's infant son Kunimatsu and executed everyone who had defended Osaka castle in 1615; the heads of the losing samurai were stuck on planks of wood all along the road from Kyoto to Fushimi. Even though he thought his 19-year-old son Tokugawa Nobuyasu innocent of plotting against Oda Nobunaga, he had to accept his death and allowed the youth to commit seppuku rather than jeopardize his own long-term political ambitions. Ieyasu always had the reputation of being a cold fish: he was never liked and never popular but feared and respected. Some have attributed his utter ruthlessness to an uncertain and dysfunctional childhood, and it is true that the trauma of his early years always stayed with him: when he became master of Japan he executed a man who had insulted him 40 years earlier. One of many quotes attributed to Ieyasu, meant to show his merciless pragmatism, is 'Life means that I can live to see tomorrow.'

In his public mode Ieyasu was a mixture of boldness and circumspection, calculation and subtlety, apparent constancy and loyalty shot through with cunning and machiavellianism. He trimmed, switched alliances, made and unmade coalitions depending on where the greatest power was at any moment. He never made the mistake of opposing Nobunaga and Hideyoshi while they were still alive. In many ways Ieyasu was an archetypal Japanese ruler, hailed by many historians as the greatest figure in Nippon's history. He was intellectually curious and pored over maps of the world with Will Adams, planning voyages to Mexico, Peru and the Philippines. He conducted a personal correspondence with James I of England (another man of unorthodox sexuality). He is enigmatic because of his combination of an advanced aesthetic sense with terrible cruelty. He had a strict ethical code but thought nothing of killing, and effortlessly combined Buddhism and bushido. His dislike of foreign adventures, suspicion of the Christians and ultimate xenophobia may relate to the famous and much discussed Japanese paranoia about invasion: do not poke a stick into the snake's hole, he seemed to say, and the serpent will not come out.

Above all, Ieyasu was the supreme political operator. As a general, considered purely from the standpoint of military excellence, he cannot rank high. He was beaten badly by Takeda Shingen in 1573 and would almost certainly have lost Sekigahara if he had not already suborned vital enemy leaders and their regiments. Yet even in his military defeats, we can see a cunning mind operating. After the battle of Mikata Gahara in 1572 he made his escape by employing the famous Empty Fort stratagem. He left the gates of an abandoned castle open, but had braziers lit and drums beating. The enemy, suspecting a trap, did not dare enter the fortress until the next day, giving Ieyasu time to escape. Ieyasu was the supreme watchmaker, the sublime chess-player, always calculating several moves ahead. He did not join in Hideyoshi's subjugation of western Japan or his invasion of China, correctly calculating that his troops would one day be needed to fight Hideyoshi's armies and should not be squandered meanwhile on the great man's quixotic ventures.

Ieyasu believed in acquiring as much knowledge as possible about as many subjects as possible, in making a close cost-benefit analysis of every decision and in trying to perceive what could not be seen by normal means – and hence the attention he paid to minute details and apparent trifles (a mindset we associate in fiction with Sherlock Holmes). His was a superior intellect to Nobunaga and Hideyoshi. Where Nobunaga was harsh and almost demonic – his detractors said he harboured designs to kill the emperor and found his own dynasty (though professional historians scoff at the idea) – and Hideyoshi was headstrong and arrogant (a Japanese Attila, perhaps), Ieyasu was devious, calculating, indirect and wide-circling. Although

the words are not authentic, two sayings attributed to Hideyoshi and
Ieyasu sum up the difference. Hideyoshi allegedly composed a haiku
as follows: 'If a bird doesn't sing, I'll make it sing.' Ieyasu was then
said to have written his own verse as a rebuke to this: 'If a bird does-
n't sing, I'll wait for it to sing.'

Ieyasu believed, like his friend the great swordsman Miyamoto
Musashi, that the mental and material worlds were all one. The celebrated military theorist Clausewitz said that war was the continuation of politics by other means, but Musashi thought that
swordsmanship was the continuation of philosophy by other means,
and Ieyasu agreed with him. There was in Ieyasu a pronounced love
of the esoteric, which partly explains his notorious closeness to the
Ninja, Japan's equivalent of the Assassins. According to their own
lights, the Ninja were not simple hitmen but the glorious flowering
of an entire outlook on life that involved the deep study of psychology and hypnotism; Ninja were supposed to have techniques that
could induce mass hysteria. One of Ieyasu's most trusted warlords
was Kattori Kanzo, the head of the Ninja. Ieyasu's preference for the
indirect and secret ways of proceeding rather than the overt tri-
umphalism of Nobunaga and Hideyoshi gives us an important clue
to the way his mind worked. He liked to use techniques of disinfor-
mation, sabotage and betrayal, reserving military force as the
weapon of last resort; this is quite clear in the long, patient campaign
to weaken Osaka castle before he finally conceded (in 1615) that
only force would work. It also explains how Ieyasu could see
through the designs of the Jesuits and the Spanish where Nobunaga
could not, though admittedly the intelligence he received from Will
Adams was important.

Ieyasu seems to have intuited that the true nature of the Jesuits
was not unlike that of his allies, the Ninja. He also believed that wisdom and knowledge could be found in the most unexpected places,
in a humble village as much as in Osaka castle. He defended his

favourite pastime – hawking – along these lines. His credo regarding the sport has been reported as follows: 'When you go into the country hawking, you learn to understand the military spirit and also the hard life of the lower classes. You exercise your muscles and train your limbs. You do any amount of running and walking and become quite indifferent to heat and cold so you are unlikely to suffer from any illness.'

Men with Ieyasu's chess-playing brilliance and patience are rare. He, even more than Napoleon, is the great example of the politician as general and war leader. Instructive historical analogies are hard to find, and perhaps only the devious, circling Franklin D. Roosevelt comes close, albeit from an entirely different culture. As an ambitious young man, Ieyasu was able to foresee the ways in which, after many a summer, he might ascend to supreme power. It would be a long, hard slog, requiring luck and patience. As a believer in Buddhist karma, Ieyasu would not have spent as much time worrying about luck as Napoleon would later, so it was the virtue of patience on which he concentrated. The aphoristic wisdom that has come down to us from Ieyasu makes him seem almost an oriental Marcus Aurelius in his embrace of stoicism. So: life is like a long journey with a heavy burden; walk slowly and make sure you don't trip; don't despair or be unhappy, for imperfection is the human lot; practise forbearance and patience; always remember the bad times you have had; one has to experience both military victory and defeat to be truly wise; find fault with yourself rather than with others; all great men understand the importance of patience. Above all else, Ieyasu emphasized patience : 'Strength means patience, which means conquering one's first impulses. There are seven emotions – joy, anger, anxiety, love, grief, fear and hate. You must resist all these to be called patient. Whatever my other weaknesses, I have known how to practise patience. My descendants will only be like me if they practise patience.'

RICHARD THE LIONHEART: A PLAIN MAN OF WAR

After the complexity of Ieyasu, the psychology of a plain man of war like Richard the Lionheart seems fairly straightforward. Richard, though no fool, was almost entirely a military figure; the great political talent of his era in western Europe was his lifelong enemy King Philip Augustus of France. Richard was one of those relatively rare characters – the man who lives for war. Yet the professional warrior who seeks to make war at all times, not for money, power, ultimate political goals or to make his name or even because he has no choice, but simply for love of combat, is in himself an enigma. In the modern era General Phil Sheridan of American Civil War fame and General George Patton are two more of the breed. Such men can puzzle even themselves: Patton famously thought, in all seriousness, that he must be a reincarnation of previous warriors. Richard was a hard, ruthless man with few scruples, a very talented general but a morally ambivalent character. The troubadour Bertran de Born called him 'Richard Yea and Nay', which is itself ambiguous. It is usually assumed that the accolade referred to his self-confidence and decisiveness: he was a man who saw things in black and white. But it could also mean that he was a man of contradictions.

The most obvious contradiction was that this professional warrior who spent 26 years in continual warfare, the ultra macho figure as it were, had also drunk deeply from the well of Aquitanian culture. Influenced by his mother, Richard absorbed through his pores the 'feminine' sensibility of southern France, the land of troubadours, minstrels, *aubade*, chivalry and the worship of women. He was by common consent a highly talented composer of romantic songs, which is the origin of the legend of his association with the troubadour Blondel. And he could be very witty. The preacher Fulk of Neuilly once sought to upbraid him by pointing out that he could never receive the grace of God while he kept at his side the three daughters Pride, Lechery and Covetousness. Richard made plain his contempt for the priestly class with his devastating reply: 'I have already given these daughters away in marriage. Pride I gave to the Templars, Lechery I gave to the Benedictines, and Covetousness to the Cistercians.'

It is sometimes said that Richard and Cortés are natural pairs as warriors, as both rationalized looting and conquest under the cloak of religion. But Richard was a deeply cynical man and had no real use for religion except as an instrument of social control. His cynicism was widely reported. When accused of using England simply as a cash cow to raise money for the crusades, he replied: 'I would have sold London itself if I could have found a buyer.' His brother John was obviously guilty of high treason for plotting to become king of England himself while Richard was away at the crusades. But when Richard returned to England after being kidnapped in Austria, he made short work of his enemies. At Lisieux in Normandy in May 1194 a terrified John threw himself on Richard's mercy. Evincing the most lofty contempt for his brother, Richard cut short the flow of insincere apologies: 'Think no more of it, John: you are only a child who has had evil counsellors.'

While stopping short of John's full-blooded atheism, Richard had no time for organized religion. In this he was following in the tradition of his Angevin father, Henry II, and his brothers, the so-called Devil's Brood, who respected religion only ever out of prudential motives. The ostensible aim of the crusades – recovering Jerusalem – was not really Richard's; he favoured an attack on Egypt. Moreover, he made a point of disregarding papal bans on such things as tournaments, the use of mercenaries and inter-Christian wars. When the papal legate Peter of Capua tried to insinuate the superior authority of the Pope in the matter of settling his endless wars with Philip of France and asked for the freedom of the warrior–bishop of Beauvais, Richard exploded with a signal demonstration of contempt for the papacy: 'The Pope! The Pope did not raise a finger to help me when I was in prison. And now he asks me to free a robber and arsonist who has never done me anything but harm! Get out of here, you liar, hypocrite, scoundrel, you bought-and-paid-for so-called church-man, and never let me see you again.' Interestingly, Richard's great opponent Saladin felt similar contempt for his 'spiritual superior' the Caliph in Baghdad, who likewise never lifted a finger to help him, and this found expression in a Richard-like explosion: 'As for the claims of the Caliph that I conquered Jerusalem with his army and under his banner – where were his banners and his army at the time? By God! I conquered Jerusalem with my own troops and under my own banner.'

Richard was in some ways the perfect warrior. As with all members of our sextet, he possessed enormous personal charisma. Attila, Cortés, Spartacus, Ieyasu, Richard and Napoleon all seem to have shared the salient characteristic that their mere entry on to a scene galvanized their followers and cowed their enemies and rivals with a kind of electrical charge of energy. Eloquent and persuasive – qualities he shared with Cortés and Napoleon – Richard was a master of tactics and strategy, a genius at siegecraft, an expert in logistics and supply-chain management, and could direct large-scale battles as convincingly as smaller encounters. If we except his poor judgement in allowing himself to become bogged down in the endless contest with Philip Augustus for the Vexin, there is only one obvious chink in his martial armour: his love of the gallery touch. All great commanders understand the importance of morale and the need to become a legend to their troops, but Richard carried this too far by insisting on micromanaging trivial military operations that did not require his presence. Cortés was nearly taken prisoner on several occasions in the thick of battle, but he was not endangering himself unnecessarily. Richard, though, did exactly this, and was much criticized for his folly and recklessness. He also insisted on riding into battle at the head of his troops, often in contexts where his presence at the front actually did his army a disservice, for, if he was killed or captured, the effect on morale would have been catastrophic.

Richard was a great warrior of the Middle Ages, probably the best in western Christendom about 1000–1300 but he appears at signal disadvantage when compared with a much greater figure, Genghis Khan's general Subudei, the greatest warrior of all time for his mastery of both military strategy and grand strategy. Instead of charging into battle unnecessarily, Subudei liked to sit on a hilltop, far from the carnage, directing operations with flags. Aged 65 when he headed the brilliant Mongol thrust into Poland and Hungary in 1241–2, Subudei had lost none of his energy even though he rivalled Ieyasu for corpulence. He was the all-time impresario of the slow withdrawal, luring the enemy to the perfect location for Mongol tactics; he used frozen rivers to penetrate into forested territory that would otherwise have nullified Mongol cavalry, and he used one army to screen another's flank, thus coordinating multiple armies across myriad mountain ranges. Richard was a very talented general with the right mentality for a warrior, but he stopped short of being an undisputed military genius.

CORTÉS: THE WARRIOR AS GAMBLER AND PLUNDERER

When it comes to evaluating Cortés and his mental processes, a central issue is the core one: just how good was he as a captain? The obvious comparison is with Francisco Pizarro, the conqueror of Peru, who achieved his success by methods remarkably similar to Cortés's: dramatic shows of force, ruthless massacres, abduction of rulers (with Atahualpa, the Inca king, playing the role of Montezuma). Cortés has always been favoured by historians, simply because he was literate and educated whereas Pizarro was more obviously a thug, illiterate and uncultivated. It has been alleged that literacy was a key factor in the conquest of the Americas, conferring a signal advantage on the conquistadores: Pizarro, for example, had read all about Cortés's exploits and could copy his methods, whereas Atahualpa knew nothing of the fate of Montezuma. But this assumes that Cortés was a true original. In fact all the typically Cortésian modes – cowing people by terror, ritual shows of force, abducting rulers – were not original to him but had been extensively used by the Spanish even before the discovery of the New World, as in the final campaigns to expel the Moors from the Iberian peninsula. It is also claimed that Cortés had a tougher task, as the Aztecs were a rising power not yet at their zenith, whereas the Incas had passed their peak and were already on the wane when Pizarro came upon them.

A sober estimate of the exploits of Cortés and Pizarro might actually end by awarding the palm to the latter. He had a smaller army (albeit better equipped) and fewer resources than Cortés, did not rely so heavily on indigenous allies and was operating at a far greater distance from his effective base, absurdly remote from any Spanish outposts that could support or reinforce him. Against that, in turn, it could be argued that the master weapon of disease, if we can so term it, was much more of an ally for the conquistadores in Peru than in Mexico. Cortés was always uneasy about the way Pizarro's feats might steal his own thunder. The two men are said to have met just once, rather like Nelson and Wellington, and in similar circumstances, as the one man was returning from his exploits, and the other on the point of setting out on his. The encounter was in Palos, Spain, in May 1528, and was especially galling for Cortés as he had earlier been expressly forbidden to attempt the conquest of Peru. Nettled by Pizarro's rival reputation, which seemed to diminish his own glory, Cortés had his secretary invent the story that Pizarro, as a foundling, had been raised by swine.

Napoleon had no doubt that Cortés was one of the great achiev-
ers of the ages and that it was humbug to try to belittle his colossal
achievements – but then Napoleon had his own reasons for wishing
to promote the myth of the great man. Some historians have tried to
demote Cortés and make him the lucky beneficiary of an innate
western superiority over the Amerindians. Thus it is variously
asserted that he won because the Aztecs were terrified of horses, that
they regarded the Spanish as returning gods, that firearms gave him
an overwhelming advantage or that the conquistadores were largely
immune to the diseases imported from Europe that devastated the
native population of the Americas. Most of these arguments fail to
convince; this is the so-called Guns, Germs and Steel thesis.

It is true that the Aztecs were initially terrified of the horse, just
as the Romans were when they first encountered elephants when
fighting king Pyrrhus of Epirus in 280–275 BC, but in both cases the
initial terror very quickly turned to familiarity and contempt. Likewise, firearms had an initial shock value but were not ultimately
significant. The 30 muskets Cortés took with him on the conquest
of Mexico were very noisy, smoky and frightening but not to
the point where they could win an empire for the Spanish. The efficiency of firearms in this era was limited by rates of firing, bullet
spin, external combustion and recoil. It should be remembered that
Aztec longbow technology was at its peak, and their arrows could
pierce steel armour that muskets could not. The critics of Cortés
are on firmer ground when they emphasize the superiority of
Toledo steel over the obsidian of the Mexica, but by the time of the
final battle for Tenochtitlan the Aztecs had acquired hundreds of
European swords and pikes from dead and sacrificed conquistadores.
As for disease, smallpox did play some part in the campaigns of
1520–1, but the really bad smallpox epidemic did not come
until 1531. In any case, the dreadful plague that finally finished
off the Aztecs in the epidemics of 1545 and 1576 was an outbreak

of Cocolitzli – a haemorrhagic fever unknown in Europe. Haemorrhagic fevers do not transmit from human to human but correlate with climatic patterns of long droughts followed by sudden rainfall.

If Cortés's reputation cannot be overthrown by the Guns, Germs and Steel thesis, there is, unfortunately for him, a crucial factor that reduces his exploits to human size. Cortés and his apologists were keen to stress the great myth of the conquest of Mexico – that it had been achieved by the heroism of a handful of intrepid men. In reality the alliance with the Tlaxcala was the key event that converted certainty of failure into near certainty of success. Some historians estimate that no less than 200,000 Tlaxcalans took part in the final Stalingrad-like destruction of Tenochtitlan. Napoleon should have reformulated his remarks in terms of his well-known conviction that what a general needs above all is luck. Cortés was lucky in that he came upon a land undergoing the convulsions of real or latent civil war. The Aztec hold on their subject peoples was precarious, with the Tlaxcalans simply waiting for the right moment to rise against them. Cortés was a catalyst, the icing on the cake of civil war in Mexico as it were. Without the Tlaxcalans, all the deterministic factors in the world – horses, mastiffs, firearms, Toledo steel, disease – would not have helped him and neither would the valour of his comrades. The best historians have recognized this. Those who like to speculate on 'what if?' counterfactual propositions in history, sometimes ask if the result of the war against the Mexica would have been different if Cortés had been killed or sacrificed after being taken prisoner (as he nearly was on a number of occasions). The overwhelming consensus is that, in alliance with the Tlaxcalans, the Spanish would still have conquered, even without Cortés; maybe the glory would have gone to Alvarado or some other leader.

All of this brings us to our most controversial proposition about the mind of Cortés – what we might term the warrior as liar. Penetrating the psychological carapace of Cortés, we can discern four main characteristics: banality, ruthlessness, the gambler's mentality and a high level of mendacity. Banal is the correct epithet to apply to a man who in many ways was no more than a glorified treasure hunter: if disease in the conquest of Mexico is a subject for investigation, we should also look at Cortés's gold fever. Given his polygamous womanizing, his lust for the yellow stuff and his yearning for titles and baubles, Cortés easily fulfils Freud's famous paradigm in answer to the question: what does a man want? Fame, fortune and the love of beautiful women was the cynical answer (Freud admitted he had nothing to say about women, who were 'the dark continent', of which he knew nothing). But the essential banality of Cortés is manifested elsewhere. In his letters to the Spanish king, when he waxes eloquent, he uses a stock of hackneyed and trite phrases, images and ideas. Some historians have been seduced into seeing Cortés's literary references as proof of his erudition and originality, but they are really just a farrago of clichés, the kind of stuff that might impress a man of limited education like Charles V. Cortés's ruthlessness hardly needs stressing after our narrative of his exploits – ruthlessness used not only against the Aztecs and other peoples but also against his fellow-Christians if they seemed to threaten his power and authority. The scruples expressed by Attila when he reproached emperor Theodosius for trying to assassinate him would have seemed merely ludicrous to Cortés, who would certainly have approved Marcian's later, successful poisoning attempt. Perhaps the most horrible and poignant single example of Cortés's murderous ruthlessness was his execution of the luckless Cuauhtémoc in 1525 on the mere suspicion of treachery, after a show trial and with complete absence of proof – an action that disgusted even his usually loyal apologist Bernal Diaz.

To fully understand the mindset of Cortés, we have to comprehend the psychology of the gambler, for his entire career is that of a man taking huge and reckless gambles. Light years of sensibility separate Cortés, pushing his luck at all points, from the careful, patient chess-player Tokugawa Ieyasu. In some ways we can align Cortés with the great voluntarists of history, those who stress willpower above all, the belief that nothing is written, nothing determined, that the so-called subjective conditions of revolutionary fervour will always trump the objective conditions of economy and society: in this sense we could put Cortés in the company of Lenin, Mao and Che Guevara. Of course the odds are stacked against voluntarists and to succeed, they must have lashings of good luck – exactly what Cortés enjoyed.

Part of his gambling mentality meant taking risks from which a more sober individual would have drawn back, but this is where we must appreciate Cortés as a true son of the Renaissance. It is no accident that we have on several occasions used the word machiavellian to describe Cortés, for both he and Machiavelli were the products of the same culture, one that glorified the warrior and held that the only true prince was the prophet armed. Although the idea has been derided in some quarters, it seems clear that the far greater self-confidence of the Renaissance and Reformation European, as compared with the Amerindians, played at least some part in the conquistadores' triumph. The gloomy world view of the Aztecs, suffused with self-doubt, with the Mexica ruled by implacable deities and forever on the brink of an Armageddon or Ragnarok, put them at a psychological disadvantage with those who believed that God was on their side, and that messianic Christianity was bound to triumph. Despite his quasi-pagan view of sexual relations and his secular pragmatism, Cortés does seem to have been genuinely devout. Even if he had not been so, his position was much more precarious than that of Richard the Lionheart, who really could afford to cock a snook at the Pope. Cortés, often on a knife-edge, had to satisfy both Charles V and the Pope that he was a loyal scion of Spanish Catholicism.

If originality is sought in Cortés, it is not in his flowery effusions to the Spanish king or in his use of tried-and-tested military techniques that it should be sought, but in his astonishing mendacity. Cortés was a liar on many different levels. He manufactured bogus documents to be sent to Spain and doctored others, hoping to wear down the bureaucrats of old Castile. He lied about the terms of his commissions, his aims and intentions and the bases on which Indian caciques submitted to him. He was a past master of the half truth, dedicated as he was to the proposition that the best way to tell a lie is to tell the truth. He claimed, falsely, that the Aztecs regarded him as a god; this myth became so persistent and was so assiduously promulgated by Christian missionaries that by the end of the sixteenth century tamed Indians were obediently parroting the myth to Catholic padres who had force-fed it to their grandfathers. Cortés's lies had both a cultural and a financial purpose. Financially, it was necessary for Cortés to promote the myth that he had conquered Mexico with a handful of devoted followers, for by providing proofs of merit he made it possible for the few (that is, his cronies) to petition the Spanish crown for titles, places and pensions. Culturally, Cortés pushed hard the notion that God allowed the conquest of the Aztecs as a favour to the Spanish monarchy and described himself as the agent of providence. As his apologist Gonzalo Fernando de Oviedo put it: 'Who can deny that the use of gunpowder against pagans is the burning of incense to Our Lord?'

In parading the big lie, Cortés had many important helpers. Both Bernal Diaz and Francisco Lopez de Gomara wrote accounts backing Cortés's self-justifying screeds at every point, portraying Cortés as the number one figure in the conquest of the Americas, with Columbus and Pizarro as lesser versions of the great man, his shadow so to speak. His achievements were particularly spun by the Franciscans, who genuinely believed that the Spanish conquest of the Americas was a prelude to the Second Coming. Needless to say,

Cortés was particularly taken with this idea. Both his secular and religious helpers, for their various reasons, colluded in Cortés's mythical view of the conquest. The Tlaxcalans and other native allies were written out, as were the bulk of the white fighting men and most of all those conquistadores who were Moorish or black. Cortés was thus successful in putting across the spectacular lie that he and a handful of paladins had defeated the Mexica unaided. It was a brilliant propaganda version of 'we few, we happy few'.

NAPOLEON: THE INTELLECTUAL WARRIOR

Where the evidence for the mind and psychology of Spartacus and Attila is tenuous, that for Napoleon is superabundant; we could easily spend an entire volume inside Napoleon's mind. He was, to begin with, by far the most intellectually gifted of our sextet. He had a range of talents rarely found in any one individual. A gifted mathematician, he also had literary flair; an outstanding general, he proved himself a consummate political plotter too, able to seize power in a *coup d'état* in France and to gauge exactly the time to do it. He combined a photographic memory with the rarer ability of total recall, and could marry complexity and lucidity in his thinking. The historian Gabriel Hanotaux spoke of 'the richest natural gifts ever received by mortal man'. Yet his failings were in some ways even greater than his talents. Although he had the intellect to have been a great chess-player like Tokugawa Ieyasu and sometimes accomplished feats that evinced such an ability, he also had the besetting sin of impatience. It was his impatience that led him to bolt his food – no meal of Napoleon's ever lasted longer than 20 minutes – but more seriously it led him to become bored and thus make policy on the wing without thinking through all the implications. It is no exaggeration to say that Napoleon brought about his own destruction by his impatience.

His attempt to strangle Britain with economic sanctions was never likely to succeed while the Royal Navy ruled the waves, but any outside chance of success depended on slow attrition over many years. As soon as the emperor realized that Russia was a weak link in his economic blockade and that the Czar was permitting exports to reach England, he made the fatal and catastrophic decision to invade Russia. Napoleon's attitude to sea power also reveals a man with no understanding of the wind and waves, forever impatient for instant solutions that might have been possible for a land army but were never practicable for a navy, above all in the age of sail with limited technology.

Impatience is an obvious character flaw, fatal in a man who would be king (or emperor). Yet in Napoleon's case we have to swim in deeper waters still. Napoleon's mind contained irrational, unintegrated motifs, which there seems no reason not to describe by the old term 'complex'. His notable misogynism probably derived from a secret (and probably justifiable) suspicion that his mother had been unfaithful before his father's early death. But the most notable 'complex' was Napoleon's fascination for all things Oriental, even when they had nothing to do with the interests of France or his empire. The obsession with the East was first noticeable on his expedition to Egypt in 1798–9. Napoleon had sold the expedition to France's executive power, the Directory, on the basis that it was desirable to acquire primary-producing areas and that a base in Egypt would threaten the British position in India. But alongside these rational motives, Napoleon was attracted by the sheer magic and mystique of the East. He told his secretary Louis Bourrienne: 'I don't want to stay here [France], there's nothing to do ... Everything's finished here but I haven't had enough glory. This tiny Europe doesn't provide enough, so I must go east ... We must go to the Orient; all great glory has been acquired there.' He loved the sights, sounds and smells of the Arab world, and felt an instinctive sympathy for the cul-

ture of the Arabs and the folkways of the sheikhs and fellahin. He
told his confidante Madame de Rémusat that he envied Alexander the
Great for having gone to India, that he liked to dress in eastern garb
and that the Orient had a special appeal to his sensibility:

> In Egypt I found myself freed from the obstacles of an irksome
> civilisation. I was full of dreams. I saw myself founding a religion,
> marching into Asia, riding an elephant, a turban on my head and in
> my hand a new Koran that I would have composed to suit my needs.
> In my undertaking I would have combined the experience of two
> worlds, exploiting for my own benefit the theatre of all history,
> attacking the power of England in India ... the time I spent in India
> was the most delightful of my life because it was the most ideal.

The integrity and security of France, French 'natural frontiers', the
defence of the French Revolution and the ideals of the Enlightenment and the *philosophes* – all these could be cited in defence of
Napoleon's constant continental warfare against the backward
monarchies of Austria and Prussia. In the case of his hostility to
Russia, we are already veering into the realm of the irrational. We
have already seen that one element in the disastrous decision to
launch the 1812 invasion was sheer impatience. But, bizarre as it
may sound, another was the 'Oriental complex'. Early in 1812
Napoleon confided to the Comte de Narbonne (an ex-royalist turned
Bonapartist) how he saw his ultimate goal if his Russian venture was
successful:

> The end of the road is India. Alexander was as far as from Moscow
> when he marched to the Ganges. I have said this to myself ever since
> St Jean d'Acre [a battle in Egypt in 1798] ... Just imagine, Moscow
> taken, Russia defeated, the Czar made over or assassinated in a palace
> plot ... and then tell me that it is impossible for a large army of

> Frenchmen and their allies to leave Tiflis and reach the Ganges.
> Essentially all that is needed is a swift stroke of a French sword for
> the entire British mercantile apparatus in the East to collapse.

Narbonne's private comment on this was: 'What a man! What ideas! What dreams! Where is the keeper of this genius? It is half-way between Bedlam and the Pantheon.' Evidently, the secret leaked out, for many in his 600,000-strong army that crossed the Vistula into Russia were not convinced that Moscow was the emperor's true aim. One officer wrote: 'Some said that Napoleon had made a secret alliance with [Czar] Alexander, and that a combined Franco-Russian army was going to march against Turkey and take hold of its possessions in Europe and Asia; others said that the war would take us to the Great Indies, to chase out the English.'

Those, like Tolstoy and others, who view Napoleon as a mere epiphenomenon, purely a creature of historical inevitability, have concentrated too much on his rational objectives. One side of Napoleon was the product of a classical sensibility, of the *philosophes* and the Enlightenment; in this sense his Promethean ambitions merely denote the intellectual who wants to possess all knowledge, the conqueror who wishes to subdue all enemies. But Napoleon's voluntarism and risk-taking, his gloom and fatalism, his love of the exotic and the abstruse and fascination with the glittering and mysterious East all come from an overdeveloped romantic imagination. The rational Napoleon despised human beings, was cynical about human nature and human motives and pessimistic about schemes to improve society. But the romantic Napoleon, in flat contradiction to these views, wanted to create a New Man and a new world. He oscillated between optimism and pessimism, at one moment saying that his marshals were actuated by a lust for glory, at another despising them for their love of money, honours, gongs and baubles.

This 'divided self' manifested itself at the political and military level in a division between rational campaigns – those of 1805–7, say, which could be justified by a French nationalist as being necessary for national security – and the highly irrational ones: Egypt in 1798–9, Russia in 1812 and, above all, Spain from 1808–13. On Spain, Napoleon played dog in the manger. He would neither agree to withdraw French armies once he saw the depth of opposition to French rule among the Spanish nor go there to command in person. Instead he left the fighting there to a set of incompetent or ill-supplied marshals, so that what became known as the Spanish ulcer continued to drain French blood and treasure for no purpose. Napoleon's irrational campaigns have always been the despair of his apologists and defenders and even of those who seek to understand him dispassionately. In despair at finding any credible interpretation, and on the Sherlock Holmes principle of 'when you have eliminated the impossible', Freud came up with two of his most far-fetched hypotheses: one that Napoleon went to Egypt out of sibling rivalry – Egypt in the Bible was the land of Joseph; the other, that guilt about divorcing Josephine as empress led Napoleon to punish himself (albeit unconsciously) in his most signal act of self-destruction when he invaded Russia in 1812.

Yet not even the world of irrationality and unintegrated complexes exhausts Napoleon's full psychological oddity. He was deeply superstitious – this is possibly the only trait that aligns him with Attila – and some have seen this as the true legacy of his Corsican upbringing. He made use of all the superstitious rites practised in Corsica: at the critical moment of a battle or at times of strong emotion he would make the sign of the cross with wide sweeps of the arm, as did the Corsican peasants of the maquis when they heard bad news. A believer in omens, portents and numerology, he disliked Fridays and the number 13 but thought certain dates were lucky for him, notably 20 March and 14 June. 'Is he lucky?' was the first question he asked about any general, before he would give him a command. He literally believed he had a lucky star and would search the heavens for his favourite dot of light at night. He famously upbraided his maternal uncle, Cardinal Joseph Fesch, when the cardinal came to protest about the emperor's treatment of Pope Pius VI. Napoleon took him outside and pointed to the night sky. 'Look up there. Do you see anything?' Fesch said he saw nothing. 'Very well,' Napoleon replied, 'in that case know when to shut up. Myself, I see my star; it is that which guides me. Don't pit your feeble and incomplete faculties against my superior organism.'

Yet even more bizarre than his belief in a lucky star was his conviction that he had a familiar spirit with whom he could converse. It is well known that patients in psychotic interludes can, as it were, exteriorize aspects of their unconscious, and in this way Napoleon constructed for himself a Little Red Man from whom he took advice. Later the bizarre incident was rehashed into a truly supernatural story whereby Napoleon encountered a genie in Egypt and made a Faustian pact with him, according to which the genie would always steer him in the right direction as long his advice was heeded. It has been hypothesized that Napoleon had several nervous breakdowns under stress that were hushed up, and it was during these unhinged periods that he had his apparent visitations from the Little Red Man.

Napoleon, in short, was a divided self, a dual man, torn between rationality and unreason, classicism and romanticism, the art of the possible and the realm of fantasy. This partly explains why he started so many hares but did not pursue them and left many projects unfinished or abandoned even though he was initially enthusiastic. First he dreamed of an empire in the western hemisphere, then abruptly abandoned the idea and sold the Louisiana territory to the USA. He signed the Concordat to ensure permanent peace with the Catholic Church, then engaged in a running battle with the papacy. From 1803–5 he was busy on a dozen different schemes for the invasion of England, which he promptly dropped after Trafalgar, as if any such idea had never entered his head. Alongside him Richard the Lionheart and Tokugawa Ieyasu seem models of sanity, but this may only be because we do not know enough about their mental life. The difficulty of applying labels like 'psychopathy' becomes immediately apparent, for whereas Attila and Cortés sometimes look like pathological cases, once again we may simply lack enough evidence of their inner lives to make a valid judgement. Certainly what is often said about Napoleon in general applies to an investigation of his psychology – that the more we learn, the less we seem to know. This is why an attempt to get inside the minds of the great warriors will always be only partially successful. Even if we were able to build up a penetrating and convincing psychological profile, it is not necessarily the case that the psychology thus revealed would tell us anything interesting about warrior capabilities.

Men of warrior calibre

The only real way we can judge warrior calibre is by reference to clearly objective criteria, which brings us back to the old trio of tactics, strategy and grand strategy. Tactics are the methods for defeating an army on a battlefield, once battle is joined, whereas strategy is the long-term design for winning the war. Most complex of all is so-called grand strategy – how to use the resources of a state, empire or bloc in warfare, on which alliances to concentrate, which parts of the war economy to emphasize and how to allocate resources between primary and secondary theatres of war. The traditional blue-water foreign policy of the United Kingdom, emphasizing the Royal Navy, is an obvious example, as was the decision by the Allies to defeat Germany first in the Second World War. History tells us that grand strategy is best directed by a statesman with a brilliant military strategist as his aide: Genghis Khan with Subudei, Abraham Lincoln with Ulysses S. Grant, Franklin D. Roosevelt with General Marshal, Ho Chi Minh with General Giap. It also appears that leaders usually come to grief when they attempt both roles: Hitler had acute geopolitical antennae but, disastrously, fancied himself also as military strategist.

The lessons hold good for our sextet of warriors. With the limitations of his personnel and their blinkered mentality, poor Spartacus could never get beyond mere tactics. Attila was a good tactician and a fair strategist but was hopeless at grand strategy. Cortés worked within limitations that prevented him from being a grand strategist in the true sense. Richard's grand strategy – to attack Saladin in Egypt – was overruled by the ideological needs of the crusade. That leaves the chess-playing Tokugawa Ieyasu as an obvious master of grand strategy. It was the tragedy of Napoleon that, although he had the intellect to be a master of grand strategy, his personality and psychology worked against this. Brilliant as he was as a tactician and strategist, he was an utter failure in his application of France's resources to his geopolitical aims.

To be true to the historical record we have sometimes had to record harsh judgements on our sextet of warriors. Yet there can be no denying that they were all, to use that debased term, 'great men'. Spartacus combined mastery of guerrilla warfare with personal charisma and the bravery born of desperation. Richard the Lionheart matched personal courage with a grasp of all aspects of military strategy, tactics and logistics; he was the perfect war-fighting technician. Attila commands reluctant admiration for his energy, the sheer force of his personality and the willingness to do whatever it took and go wherever military logic dictated. Cortés combined restless energy, superhuman willpower and a belief in his own star with a gambler's winning streak (if the wordplay can be forgiven, it might be said that he had the punter's luck in spades). Tokugawa Ieyasu was primarily a political genius who ensured that his battles were won before he fought them: he is one of history's greatest testimonies to cat-at-a-mousehole patience. Napoleon was perhaps the only incontestable military genius of the six, but his genius extended to politics, political theory and even lawmaking.

The great warrior must be a master of strategy and tactics, have high military talents, boldness, cunning, self-belief, be lucky, fight in the right circumstances and against an almost equally matched foe. On these criteria Napoleon and Ieyasu would emerge at the top of the heap, while Cortés and Spartacus, because of the second-rate opposition they faced, would rank lower down. Despite his ultimate failure, one would be inclined to rate Attila ahead of them, if only because he had to contend with at least three first-rate figures who out-thought him: Marcian, Aetius and Geiseric. Richard the Lionheart defeated the best the western and Middle Eastern world could throw against him, but just misses the first rank because of his showmanship and the gallery touch.

We can applaud the great warriors but also, curiously, we can pity them. Apart from Ieyasu, none of them died happily in bed. Richard perished stupidly because he could not leave well alone and taunted a sharpshooter. Napoleon passed away in exile on a rocky isle, almost certainly the victim of poisoning. Attila likewise died ingloriously, of poison or, if you prefer the orthodox version, in a drunken stupor. Cortés eked out his days in Spain, a forgotten if not despised figure. Only Spartacus died a hero's death in battle. But they were all titanic figures who left an indelible mark on history.

Select Bibliography

Ailes, Marianne (ed.), *Ambroise's Estoire de la Guerre Sainte*, (Woodbridge, 2003)

Ammianus, Marcellinus, *The Later Roman Empire*, (trans. W. Hamilton), (1986)

Appian, *Civil Wars*, (trans. J. Carter), (1996)

Babcock, Michael A., *The Night Attila Died*, (New York, 2005)

Baha al-Din, Ibn Shaddad, *The Rare and Excellent History of Saladin*, (trans. D.S. Richards), (Aldershot, 2001)

Barrois, C., *Psychoanalyse du Guerrier*, (Paris, 1993)

Boxer, C.R., *The Christian Century in Japan, 1549–1650*, (Berkeley, 1951)

Boyle, David, *Blondel's Song: the Capture, Imprisonment and Ransom of Richard the Lionheart*, (Viking, 2005)

Bradley, Keith R., *Slavery and Rebellion in the Roman World 140 BC–70 BC*, (Bloomington, Indiana, 1989)

Bryant, Anthony, *Sekigahara 1600: the Final Struggle for Power*, (Greenwood Press, 2005)

Chandler, David, *The Campaigns of Napoleon*, (1966)

Chandler, David, *Napoleon's Marshals*, (1987)

Clendinne, Inga, *Aztecs, An Interpretation*, (Cambridge, 1993)

Cook, Noble David, *Born to Die: Disease and the New World Conquest, 1492– 1650*, (Cambridge, 1998)

Correspondance de Napoleon, 32 vols (Paris, 1858–70)

Cortes, Hernan, Letters from Mexico, (ed. Anthony Pagden), (Yale, 1986)

Cottin, Paul, Toulon et les Anglais, (Paris, 1893)

Diamond, Jared, Guns, Germs and Steel: the Fate of Human Societies, (New York, 1997)

Diaz, Bernal, The Conquest of Mexico, (Penguin, 1963)

Doppet, F.A., Memoires, (1797)

Edbury, P.W., The Conquest of Jerusalem and the Third Crusade, (Scolar Press, 1996)

Elison, George & Bardwell Smith, L. (eds), Warlords, Artists and Commoners: Japan in the Sixteenth Century, (Honolulu, 1981)

Florus, Epitome of Roman History, (1947)

Forczyk, Robert A., Toulon 1793: Napoleon's First Great Victory, (Osprey, 2005)

Fox, C.J., Napoleon and the Siege of Toulon, (1902)

Gibbon, Edward, The Decline and Fall of the Roman Empire, (Everyman, 1993)

Gillingham, John, Richard I, (Yale, 1999)

Gordon, C.D., The Age of Attila: Fifth Century Byzantium and the Barbarians, (Ann Arbor, 1966)

Gray, J. Ellen, The Warriors: Reflections on Men in Battle, (Nebraska, 1998)

Greene, Robert, The Thirty-Three Strategies of War, (Profile, 2006)

Grousset, P., L'Empire des Steppes, (1960)

Hall, John W. & Jansen, Marius B. (eds), Studies in the Institutional History of Early Modern Japan, (Princeton, 1968)

Hazig, Ross, Mexico and the Spanish Conquest, (New York, 1994)

Heather, Peter, The Fall of the Roman Empire: a New History, (Macmillan, 2005)

Hopkins, Keith, Death and Renewal, (Cambridge, 1983)

Howarth, Patrick, Attila King of the Huns: the Man and the Myth, (Constable & Robinson, 1994)

Ireland, Bernard, The Fall of Toulon: the Last Opportunity to Defeat the French Revolution, (Weidenfeld, 2005)

Jordanes, Gothic History, (trans. C.C. Mierow), (1960)

Lamers, Jeroen, Japonius Tyrannus: the Japanese Warlord Oda Nobunaga Reconsidered, (Leiden, 2000)

Leon-Portilla, Miguel, The Broken Spears: the Aztec Account of the Conquest of Mexico, (Boston, 1992)

Leupp, Gary, *Male Colors: the Construction of Homosexuality in Tokugawa Japan,* (Berkeley, 1997)

Lyons, Malcom C. & Jackson, D.E.P., *Saladin and the Politics of Holy War,* (Cambridge, 1982)

Maenchen-Helfen, Otto, *The World of the Huns,* (California, 1993)

Man, John, *Attila: a Barbarian King and the Fall of Rome,* (Bantam Press, 2005)

Memoires du Marechal Marmont, duc de Raguse, 9 vols, (Paris, 1857)

McLynn, Frank, *Napoleon: a Biography,* (Arcade, 1997)

McLynn, Frank, *Lionheart and Lackland: King Richard, King John and the Wars of Conquest,* (Jonathan Cape, 2006)

McMullin, Neil, *Buddhism and the State in Sixteenth Century Japan,* (Princeton, 1984)

Plutarch, *Fall of the Roman Republic,* (trans. Rex Warner), (1972)

Pocock, Tom, *A Thirst for Glory: the Life of Admiral Sir Sidney Smith,* (Pimlico, 1998)

Reston, James, *Warriors of God: Richard the Lionheart and Saladin in the Third Crusade,* (Faber & Faber, 2001)

Restall, Matthew, *Seven Myths of the Spanish Conquest,* (Oxford, 2003)

Sadler, A.L., *The Maker of Modern Japan,* (1937)

Sallust, *Conspiracy of Catiline and the Wars of Jugurtha,* (1924)

Seager, Robin, *Pompey: a Political Biography,* (Oxford, 1979)

Shaw, Brent D., *Spartacus and the Slave Wars,* (Macmillan, 2001)

Sherman, Nancy, *Stoic Warriors,* (Oxford, 2005)

Smail, R.C., *Crusading Warfare,* (Cambridge, 1995)

Soustelle, Jacques, *Daily Life of the Aztecs on the Eve of the Spanish Conquest,* (1964)

Tames, Richard, *Servant of the Shogun,* (Saint Martin's Press, 1988)

Thomas, Hugh, *The Conquest of Mexico,* (Hutchinson, 1993)

Thompson, E.A., *The Huns,* (Oxford, 1996)

Totman, Conrad, *Politics in the Tokugawa Bakufu 1600–1843,* (Harvard, 1967)

Totman, Conrad, *Tokugawa Ieyasu: Shogun,* (San Francisco, 1983)

Trow, M.J., *Spartacus: the Myth and the Man,* (Sutton, 2006)

Turnbull, Stephen, *The Samurai,* (New York, 1977)

Turnbull, Stephen, *Battles of the Samurai,* (Weidenfeld, 1987)

Turnbull, Stephen, *Osaka 1614–15: the last Samurai Battle,* (Osprey, 2006)

Tyerman, Christopher, *God's War: a New History of the Crusades*, (Allen Lane, 2006)

Urbainczyk, Theresa, *Spartacus*, (Bristol Classical Press, 2004)

Ward, Allen Mason, *Marcus Crassus and the Late Roman Republic*, (Columbia, Missouri, 1977)

Winkler, Martin M., *Spartacus: Film and History*, (Blackwell, 2006)

Index

INDEX

PICTURE CREDITS

BBC Books would like to thank the following individuals and organizations for providing photographs and for permission to reproduce copyright material. While every effort has been made to trace and acknowledge copyright holders, we would like to apologize should there be any errors or omissions.

Abbreviations: t: top, b: bottom, l: left, r: right, tl: top left, tr: top right, tc: top centre, bl: bottom left, br: bottom right.

PLATE SECTION A

1t: akg-images/Peter Connolly; 1b: Charles Walker/TopFoto; 2-3tc: Werner Forman Archive/Edgar Knobloch; 3tr: Alinari Archives/Corbis; 2b: akg-images/Erich Lessing; 3b: Bryna/Universal/The Kobal Collection; 4t: The Art Archive/Galleria Nazionale della Umbria Perugia/Dagli Orti; 4b: Universal Pictures/Album/AKG; 5t: Archivo Iconografico, S.A./Corbis; 5bl: The Art Archive/Jan Vinchon Numismatist Paris/Dagli Orti; 5br: BPK/SBB; 6t: akg-images; 6b: akg-images/British Library; 7t: akg-images/Erich Lessing; 7b: British Library Images; 8t: British Library Images; 8b: Craig Aurness/Corbis.

PLATE SECTION B

1t: The Art Archive/National Archives Mexico/Mireille Vautier; 1b: The Art Archive; 2t: The Art Archive/Museo Ciudad Mexico/Dagli Orti; 2b: The Art Archive/Academia BB AA S Fernando Madrid/Dagli Orti; 3t: The Art Archive/Mireille Vautier; 3b: C20th Fox courtesy Ronald Grant Archive; 4tl: Topham Picturepoint; 4-5: Werner Forman Archive/Kuroda Collection, Japan; 6t: Werner Foreman Archive/L J Anderson Collection; 6b: Kurosawa Productions/Toho Company courtesy Ronald Grant Archive; 7t: Musée Nat. du Chateau de Malmaison, Rueil-Malmaison, France/The Bridgeman Art Library; 7b-8t: akg-images; 8b: Private Collection, The Stapleton Collection/The Bridgeman Art Library.